Photographic and Descriptive Musculoskeletal Atlas of Chimpanzees

With notes on the attachments, variations, innervation, function and synonymy and weight of the muscles

Photographic and Descriptive Musculoskeletal Atlas of Chimpanzees

With notes on the attachments, variations, innervation, function and synonymy and weight of the muscles

- Rui Diogo
- Juan F. Pastor
- Eva M. Ferrero
- Mercedes Barbosa
- Anne M. Burrows
- Bernard A. Wood

- Josep M. Potau
- Félix J. de Paz
- Gaëlle Bello
- M. Ashraf Aziz
- Julia Arias-Martorell

CRC Press
Taylor & Francis Group
Boca Raton London New York

CRC Press is an imprint of the
Taylor & Francis Group, an **informa** business

A SCIENCE PUBLISHERS BOOK

CRC Press
Taylor & Francis Group
6000 Broken Sound Parkway NW, Suite 300
Boca Raton, FL 33487-2742

First issued in paperback 2019

© 2013 by Taylor & Francis Group, LLC
CRC Press is an imprint of Taylor & Francis Group, an Informa business

No claim to original U.S. Government works

ISBN-13: 978-1-4665-8018-3 (hbk)
ISBN-13: 978-0-367-38035-9 (pbk)

This book contains information obtained from authentic and highly regarded sources. Reasonable efforts have been made to publish reliable data and information, but the author and publisher cannot assume responsibility for the validity of all materials or the consequences of their use. The authors and publishers have attempted to trace the copyright holders of all material reproduced in this publication and apologize to copyright holders if permission to publish in this form has not been obtained. If any copyright material has not been acknowledged please write and let us know so we may rectify in any future reprint.

Except as permitted under U.S. Copyright Law, no part of this book may be reprinted, reproduced, transmitted, or utilized in any form by any electronic, mechanical, or other means, now known or hereafter invented, including photocopying, microfilming, and recording, or in any information storage or retrieval system, without written permission from the publishers.

For permission to photocopy or use material electronically from this work, please access www.copyright.com (http://www.copyright.com/) or contact the Copyright Clearance Center, Inc. (CCC), 222 Rosewood Drive, Danvers, MA 01923, 978-750-8400. CCC is a not-for-profit organization that provides licenses and registration for a variety of users. For organizations that have been granted a photocopy license by the CCC, a separate system of payment has been arranged.

Trademark Notice: Product or corporate names may be trademarks or registered trademarks, and are used only for identification and explanation without intent to infringe.

**Visit the Taylor & Francis Web site at
http://www.taylorandfrancis.com**

**and the CRC Press Web site at
http://www.crcpress.com**

Preface

Chimpanzees, including common chimpanzees (*Pan troglodytes*) and bonobos (*Pan troglodytes*), are our closest living relatives. This photographic and descriptive musculoskeletal atlas of *Pan* follows the organization used in the photographic atlases of *Gorilla* and *Hylobates* published by Diogo et al. in 2010 and 2012, respectively. These three books are part of a monograph series that will be completed by an atlas of orangutans. The series is designed to help provide the comparative, phylogenetic, and evolutionary context for understanding the evolutionary history of the gross anatomy of modern humans and our closest relatives.

We dissected and took high-quality photographs of 12 specimens of common chimpanzees, including infants and adults and both males and females (see Methodology and Material below) and for one of these specimens (VU PT1) we were able to record the wet weight of many of the muscles. Where there are differences between the myology (e.g., the presence/absence of a muscle or a muscle bundle, its attachments and/or its innervation) of this specimen and any of the other specimens dissected by us, we provide detailed comparative notes, and use photographs to document the differences.

The atlas also includes the results of an extensive review of the literature about the musculature of both common chimpanzees and bonobos, a comprehensive review of muscle variants among individual members of these two species, and a list of the synonyms used in the literature to refer to the muscles of these primates. The data previously obtained from our dissections of numerous primates and other mammalian and non-mammalian vertebrates (e.g., Diogo 2004a,b, 2007, 2008, 2009, Diogo & Abdala 2007, 2010, Diogo et al. 2008, 2009a,b, 2010, 2012, Diogo & Wood 2011, 2012) were used to test hypotheses about the homologies among the muscles of chimpanzees, other apes and modern humans, and other taxa.

We hope this atlas will be of interest to students, teachers and researchers studying primatology, comparative anatomy, functional morphology, zoology, and physical anthropology, as well as to clinicians and researchers who are interested in understanding the origin, evolution, and homology of the musculoskeletal system of modern humans as well as the comparative context of common variants on the musculature of modern humans.

Acknowledgements

We gratefully acknowledge support and funding from all of the institutions and funding bodies that made this project possible; we especially acknowledge support for RD's research via a GW Presidential Fellowship and a Howard University College of Medicine start-up package. We particularly acknowledge B. Richmond (George Washington University) for allowing us to dissect the GWUANT PT1 and GWUANT PT2 specimens, J. Fritz and J. Murphy (Primate Foundation of Arizona) for providing the PFA 1016, PFA 1009, PFA 1051, PFA 1077 and PFA UNC specimens, the Fundacion Mona for providing the VU PT1 specimen, the Zoo-Aquarium of Madrid for providing the VU PT2 specimen, the Bioparc Fuengirola for providing the VU PT3 specimen, and Lisa Parr (Emory University) for providing the Yerkes uncatalogued specimen.

Contents

Preface	v
Acknowledgements	vii
1. Methodology and Material	**1**
2. Head and Neck Musculature	**4**
3.1 Mandibular musculature	4
3.2 Hyoid musculature	6
3.3 Branchial musculature	19
3.4 Hypobranchial musculature	27
3.5 Extra-ocular musculature	30
3. Pectoral and Upper Limb Musculature	**31**
4. Trunk and Back Musculature	**63**
5. Diaphragmatic and Abdominal Musculature	**69**
6. Perineal, Coccygeal and Anal Musculature	**73**
7. Pelvic and Lower Limb Musculature	**76**
Appendix I: Literature Including Information about the Muscles of Chimpanzees	93
Appendix II: Literature Cited, Not Including Information about the Muscles of Chimpanzees	105
Index	107
About The Authors	113
Color Plate Section	**115**

CHAPTER 1

Methodology and Material

The 12 common chimpanzees (*Pan troglodytes*) dissected for this study were made available by the following institutions: George Washington University (GWUANT PT1 and GWUANT PT2, adult females, formalin embalmed), Howard University (HU PT1, infant male, formalin embalmed), Duquesne University (Yerkes Regional Primate Center, uncatalogued adult male, formalin embalmed), Valladolid University (VU PT1, 81 kilograms, adult male, fresh, provided by the Fundacion Mona; VU PT2, 72 kilograms, adult male, fresh, provided by Zoo-Aquarium of Madrid; VU PT3, 58 kilograms, adult female, fresh, provided by the Bioparc Fuengirola), and Primate Foundation of Arizona (PFA 1016, 50-years old adult female, fresh; PFA 1009, adult female, fresh; PFA 1051, 1-year old infant female, fresh; PFA 1077, infant female, fresh; PFA UNC—uncatalogued—infant male, fresh). We took photographs of the musculoskeletal system of all of the specimens we dissected, but the muscle weights listed in this atlas are from the VU PT1 specimen (total body weight = 81 kg) that was in particularly good condition. The photographs of the osteological structures shown in this atlas are from the VU PT3 specimen. In the text below, when the data are available, we provide for each muscle: 1) its weight in the VU PT1 specimen. The total mass of all the striated muscles is given in parentheses immediately following the name of the muscle. In the case of paired muscles, the muscles of the left and right sides are referred to as LSB and RSB, respectively. When the muscle is part of a symmetrical structure (e.g., stylohyoideus), the weight given is that of the muscle of one side of the body; when the muscle is unpaired (e.g., diaphragm), the weight given is that of the part of the muscle that comes from one side only; 2) the most common attachments and innervation of the muscle within the chimpanzee clade, based on our dissections and on our literature review; 3) the function of the muscle (e.g., based on electromyographic studies—EMG—of neck and upper limb muscles or the use of the Facial Action Coding System—FACS—for facial muscles); 4) comparative notes, especially when where there are differences (e.g., regarding the presence/absence of the muscle, or of its bundles, its attachments, and/or its innervation) between the configuration usually found in chimpanzeess and the configuration found in a specimen dissected by us (in these cases we often provide photographs to illustrate the differences) or by others; and 5) a list of the synonyms that have been used by other authors to designate that muscle.

Apart from the chimpanzee specimens mentioned above, we have dissected numerous specimens from most vertebrate groups, including bony fish, amphibians, reptiles, monotremes, rodents, colugos, tree-shrews, and numerous primates, including modern humans (a complete list of these specimens and the terminology used to describe them is given in Diogo & Abdala 2010 and Diogo & Wood 2011, 2012). This broad comparative context proved to be crucial for generating hypotheses about the homologies among the muscular structures of hylobatids, modern humans and other primate and non-primate vertebrates, and it also informed the nomenclature we proposed by Diogo et al. (2008, 2009a,b), Diogo & Abdala (2010) and Diogo & Wood (2011, 2012). This nomenclature is based on that employed in modern human anatomy (e.g., Terminologia Anatomica 1998), but it also takes into account the names used by researchers who have focused on non-human mammals (e.g., Saban 1968, Jouffroy 1971). In the majority of the figures we use Latin names for the soft tissues and anglicized names for the bones. In the figures that mainly illustrate osteological structures we use Latin names, but to avoid redundancy when these names are similar to the anglicized version (e.g., processus mastoideus = mastoid process) we do not provide the latter; in those cases in which they are substantially different (e.g., incisura mandibulae and mandibular notch) we provide both the Latin names and the anglicized version.

When we describe the position, attachments and orientation of the muscles and we use the terms anterior, posterior, dorsal and ventral in the sense in which those terms are applied to pronograde tetrapods (e.g., the sternohyoideus mainly runs from the sternum, posteriorly, to the hyoid bone, anteriorly, and passes mainly ventrally to the larynx, which is, in turn, ventral to the esophagus; the flexors of the forearm are mainly situated on the ventral side of the forearm). However, the nomenclature used in Terminologia Anatomica (1998) was defined on the basis of an upright posture and although most primates are not bipeds, nearly all of the osteological names and most of the myological ones used by other authors (and by us) to designate the structures of non-human primates, including chimpanzees, follow the Terminologia Anatomica nomenclature. Although this is potentially confusing we judged it to be preferable to refer to the topology of the musculoskeletal structures of non-human primates in this way because in the vast majority of primates the 'superior angle of the scapula' is actually mainly anterior, and not superior, to the 'inferior angle of the scapula', and the 'cricoarytenoideus posterior' actually lies more on the dorsal, and not on the posterior, surface of the larynx. Moreover, we think that, by keeping in mind that the actual names (both in Latin and in English) of all the osteological structures and of most the myological structures mentioned in this atlas refer to a biped posture while the actual descriptions provided here regarding the topology of these structures refer to a pronograde posture, most readers will have no difficulty in interpreting and understanding the information provided in this book.

The muscles listed below are those that are usually present in adult chimpanzees; muscles that are only occasionally present in adult chimpanzees are discussed in other parts of the atlas. In our written descriptions, we follow Edgeworth (1935), Diogo & Abdala (2010) and Diogo & Wood (2011, 2012) and divide the head and neck muscles in five main subgroups: 1) mandibular, muscles that are generally innervated by the fifth cranial nerve (CN5) and include the masticatory muscles, among others; 2) hyoid, muscles that are usually innervated by CN7 and include the facial muscles, among others; 3) branchial, muscles that are usually innervated by CNC9 and CN10, and include most laryngeal and pharyngeal muscles, among others; 4) hypobranchial, muscles that include all the infrahyoid and tongue muscles, and the geniohyoideus. According to Edgeworth (1935) the hypobranchial muscles were developed primarily from the anterior myotomes of the body and then migrated into the head. Although they retain a main innervation from spinal nerves, they may also be innervated by CN11 and CN12, but they usually do not receive any branches from CN5, CN6, CN7, CN8, CN9 and CN10; 5) extra-ocular, muscles that are usually innervated by nerves CN3, CN4 and/or CN6 in vertebrates. The head, neck, pectoral and upper limb muscles are listed following the order used by Diogo et al. (2008, 2009a), Diogo & Abdala (2010) and Diogo & Wood (2011, 2012), while the pelvic and lower limb muscles, as well as the other muscles of the body, are listed following the order used by Gibbs (1999). It should be emphasized that the literature review undertaken by this latter author provided a crucial basis and contribution for our own literature review.

CHAPTER 2

Head and Neck Musculature

3.1 Mandibular musculature

Mylohyoideus (Figs. 7–11)
- Usual attachments: From the mylohyoid line of the mandible to the hyoid bone, posteriorly, and to the ventral midline, anteriorly.
- Usual innervation: Mylohyoid nerve of mandibular division of CN5 (Miller 1952: *P. paniscus*).
- Notes: There is usually no distinct median raphe of the mylohyoideus in chimpanzees, according to Swindler & Wood (1973) and to our dissections. However, Göllner (1982) reported a median raphe in the two adult *P. troglodytes* dissected by him. The **intermandibularis anterior** (present in some mammals) is not present as a distinct muscle in chimpanzees.
- Synonymy: Intermandibularis (Edgeworth 1935).

Digastricus anterior (Figs. 7–11, 18–19)
- Usual attachments: From the intermediate tendon of the digastric, and sometimes also from the hyoid bone (e.g., Gratiolet & Alix 1866, Miller 1952), to the mandible; it usually meets its counterpart in the midline (e.g., Vrolik 1841, Wilder 1862, Sonntag 1923, Miller 1952, Dubrul 1958, Starck & Schneider 1960, Göllner 1982, our dissections).
- Usual innervation: Mylohyoid nerve of mandibular division of CN5 (Miller 1952: *P. paniscus*).
- Synonymy: Anterior belly of digastricus (Miller 1952); anterior belly of biventer mandibulae (Starck & Schneider 1960).

Tensor tympani
- Usual attachments: From the auditory tube and adjacent regions of the neurocranium to the manubrium of the malleus.
- Usual innervation: Data are not available.
- Notes: According to Maier (2008) in chimpanzees the chorda tympani usually passes above the tensor tympani (i.e., the epitensoric condition).

Tensor veli palatini (Figs. 14–15)
- Usual attachments: From the Eustachian tube and the adjacent regions of the cranium (usually the scaphoid fossa of the medial pterygoid plate: Gratiolet & Alix 1866, Sonntag 1923, 1924, Dean 1985; sometimes the apex of the petrous temporal bone: e.g., Sonntag 1923) to the pterygoid hamulus and soft palate.
- Usual innervation: Data are not available.
- Notes: In modern human infants and in adult apes, including chimpanzees, the palate lies much closer to the roof of the nasopharynx than it usually does in adult modern humans, so in the former the levator veli palatini and tensor veli palatini do not run so markedly downwards to reach the palate as they do in the latter. The **pterygotympanicus** (present, e.g., as an anomaly in modern humans) is usually missing in chimpanzees.
- Synonymy: Peristaphylin externe (Gratiolet & Alix 1866); tensor palatini (Sonntag 1923).

Masseter (Figs. 3, 8–12, 18, 20)
- Usual attachments: Mainly from the zygomatic arch; the pars superficialis inserts mainly onto the lower edge of the base of the mandible, while the pars profunda inserts mainly onto the ascending ramus and the coronoid process of the mandible.
- Usual innervation: Branch of the mandibular division of CN5 (Miller 1952: *P. paniscus*).
- Notes: In chimpanzees there is often a strong fascia/aponeurosis between the superficial and deep heads of the masseter (e.g., Sonntag 1923, Miller 1952). The **zygomatico-mandibularis** is usually not present as a separate muscle in chimpanzees, but it is present as a separate bundle of the masseter in the *P. troglodytes* specimens dissected by Göllner (1982) and Gratiolet & Alix (1866) and embryo dissected by Starck (1973) and at least some of the chimpanzees dissected by us (PFA 1016, PFA 1009, PFA 1051, HU PT1), lying superficially to the temporalis and running from the deep surface of the zygomatic process to the coronoid process of the mandible.

Temporalis (Figs. 3, 8–11, 20)
- Usual attachments: From the whole of the fossa temporalis, the temporalis fascia and adjacent regions of the cranium to the coronoid process and ramus of the mandible.
- Usual innervation: Branch of the mandibular division of CN5 (Miller 1952: *P. paniscus*).
- Notes: Swindler & Wood (1973) stated that in chimpanzees the temporalis is usually not divided into a distinct pars superficialis and a distinct pars profunda. However Göllner (1982) dissected 4 neonate and 2 adult *P. troglodytes* and found a temporalis divided into a pars superficialis and a pars profunda in these 6 specimens. In the specimens dissected by us there was in general no notable

differentiation of the temporalis into a pars superficialis and a pars profunda, but in the specimens PFA 1077 and PFA UNC there was some differentiation between the anterior, more vertical, and deeper fibers of the temporalis and the posterior, more oblique, and superficial fibers of this muscle. Regarding the pars suprazygomatica, in some neonates of *P. troglodytes* dissected by Göllner (1982) and by us this part of the muscle is clearly differentiated, but in the adult chimpanzees dissected by us and by other authors, including Göllner (1982), the pars suprazygomatica is usually not present as a distinct bundle.

Pterygoideus lateralis (Fig. 13)
- Usual attachments: From the lateral pterygoid plate and the adjacent regions of the cranium (e.g., great wing of the sphenoid bone: Miller 1952) to the capsule of the temporomandibular joint and the neck of the mandibular condyle.
- Usual innervation: Branch of the mandibular division of CN5 (Miller 1952: *P. paniscus*).
- Notes: Göllner (1982) suggested that the inferior and superior heads of the pterygoideus lateralis are are not differentiated in the neonate and adult *P. troglodytes* dissected by him, but the two heads were clearly present in all the *Pan* specimens dissected by Gratiolet & Alix (1866), Sonntag (1923), Miller (1952), and by us (including neonates and adults).
- Synonymy: Pterygoideus externus (Gratiolet & Alix 1866, Sonntag 1923, Miller 1952).

Pterygoideus medialis (Figs. 7, 12–13, 18)
- Usual attachments: Mainly from the medial surface of the lateral pterygoid plate of the sphenoid bone and adjacent regions of the skull (e.g., Miller 1952) to the medial side of the mandible.
- Usual innervation: Branch of the mandibular division of CN5 (Miller 1952: *P. paniscus*).
- Synonymy: Pterygoideus internus (Gratiolet & Alix 1866, Sonntag 1923, Miller 1952).

3.2 Hyoid musculature

Stylohyoideus (Figs. 7–10)
- Usual attachments: From the styloid process and/or adjacent regions of the skull to the hyoid bone.
- Usual innervation: Branch of CN7 (Miller 1952: *P. paniscus*).
- Notes: A piercing of the styhyoideus by the intermediate digastric tendon was found in the *P. troglodytes* specimens dissected by Sonntag (1923) and in most chimpanzees dissected by us, but in our HU PT1 specimen and in the bonobo dissected by Miller (1952) the stylohyoideus was mainly superficial to the posterior digastricus. To our knowledge the **jugulohyoideus** (usually present

in strepsirrhines and sometimes present in *Tarsius*) and the **stylolaryngeus** (sometimes present in orangutans) have never been reported in chimpanzees.

Digastricus posterior (Figs. 7–10, 18)
- Usual attachments: From the mastoid region and adjacent regions of the skull (e.g., sometimes from the occipital bone: Tyson 1699) to the digastric intermediate tendon.
- Usual innervation: Branch of CN7 (Miller 1952: *P. paniscus*).
- Synonymy: Posterior belly of digastricus (Miller 1952); posterior part of biventer mandibulae (Starck & Schneider 1960).

Stapedius
- Usual attachments: Probably inserts onto the stapes, but the information provided in the literature is sparse.
- Usual innervation: Data are not available.

Platysma myoides (Figs. 1–3)
- Usual attachments: Main from the pectoral region and the neck to the modiolus and adjacent regions of the mouth.
- Usual innervation: 'Platysma' (platysma myoides *sensu* the present study) by cervical branch of CN7; 'transversus nuchae' (remaining of platysma cervicale *sensu* the present study) by posterior auricular branch of CN7 (Miller 1952: *P. paniscus*).
- Notes: The **platysma cervicale**, **sphincter colli superficialis** and **sphincter colli profundus** are usually not present as distinct muscles in chimpanzees. Influential authors such as Owen (1830-1831) and Sonntag (1924) used 'platysma myoides' to describe the platysma complex of Asian apes such as orangutans, and this nomenclature has been followed by various researchers, including Seiler (1976), and was thus also followed in Diogo et al.'s (2009b) review. However, Owen (1830-1831) stated that the 'platysma myoides' of orangutans incorporates the platysma myoides of modern humans plus the platysma cervicale of other mammals, and our recent dissections of numerous primates and comparisons with the data provided in the literature confirm that orangutans and hylobatids usually have a platysma cervicale (i.e., a part of the platysma complex that attaches onto the nuchal region). That is, juvenile and adult orangutans and hylobatids usually have a well-developed platysma cervicale similar to the muscle that is found in most other primates and that is usually markedly reduced, or even absent, in *Pan* (including the neonates dissected by us), *Gorilla* and *Homo*. In fact, juveniles and adult members of the genus *Pan* usually do not have a well-developed platysma cervicale, often having instead a small muscle 'transversus nuchae' (e.g., Gratiolet & Alix 1866, Broca 1869, Champneys 1872, Macalister 1871, Chapman 1879, Sutton 1883, Virchow 1915, Sonntag 1923, Sullivan & Osgood 1925, Huber 1930b, 1931, Loth 1931, Miller 1952, Swindler & Wood 1973, Pellatt 1979, our dissections,

e.g., Fig. 1 of plate IX of Gratiolet & Alix 1866, Fig. 5 of Sullivan & Osgood 1925, Fig. 35 of Huber 1930b, Fig. 11 of Huber 1931, Fig. 785 of Edgeworth 1935, Fig. 12 of Miller 1952). This 'transversus nuchae' seems to be a vestige of the platysma cervicale (e.g., Loth 1931, Diogo & Wood 2011, 2012). Regarding the sphincter colli superficialis, it should be noted that the 'sphincter colli' described by Burrows et al. (2006) in chimpanzees could correspond to the sphincter colli superficialis *sensu* the present work, because they stated that this structure is superficial to the platysma myoides. However, the sphincter colli superficialis is not present as a distinct muscle in all the other *Pan* specimens described in the literature and dissected by us and, as explained above, is actually usually missing in all extant primate taxa. To our knowledge, the sphincter colli profundus has never been described in *Pan*.
- Synonymy: Peaucier propre plus risorius (Gratiolet & Alix 1866); platysma (MacAlister 1871, Miller 1952); tracheloplatysma (Swindler & Wood 1973).

Occipitalis (Fig. 5)
- Usual attachments: Mainly from the occipital region to the galea aponeurotica.
- Usual innervation: Posterior auricular branch of CN7 (Miller 1952: *P. paniscus*).
- Notes: In hylobatids and non-hominoid primates the occipitalis is usually differentiated into a main body (or 'occipitalis proprius') and a 'cervico-auriculo-occipitalis' (*sensu* Lightoller 1925, 1928, 1934, 1939, 1940a,b, 1942). The latter is a lateral/superficial bundle of the occipitalis that often runs anterolaterally from the occipital region to the posterior portion of the ear and that sometimes covers part of the auricularis posterior in lateral view. The 'cervico-auriculo-occipitalis' of primates such as *Macaca* was designated as the 'deep layer of the occipitalis' by Huber (1930b, 1931), but it is not homologous to the 'pars profunda' found in *Pongo* by Sullivan & Osgood (1925) and by us, nor to the 'pars profunda' described in *Pan* by Burrows et al. (2006), nor to the 'pars profunda' *sensu* Seiler (1976), which corresponds to the 'occipitalis proprius' *sensu* the present study. Our dissections and the literature reviewed by us (for detailed accounts on this issue see, e.g., Lightoller 1928a, 1934, Seiler 1976) confirm that the 'cervico-auriculo-occipitalis' is usually not present as a distinct structure in *Pan*, although it was apparently present as a small structure in a chimpanzee illustrated by Seiler (1976: see his Fig. 143).
- Synonymy: Part of occipito-frontalis (Owen 18368, Macalister 1871, Sutton 1883).

Auricularis posterior
- Usual attachments: From the occipital region to the posterior region of the ear.

- Usual innervation: Posterior auricular branch of CN7 (Miller 1952: *P. paniscus*).
- Notes: The auricularis posterior of chimpanzees is usually undivided but in the *P. troglodytes* specimens dissected by authors such as MacAlister (1871) and Pellatt (1979) the muscle had two bundles.
- Synonymy: Retrahens aurem (Wilder 1862, Macalister 1871); attrahens aurem plus retrahens aurem (Sonntag 1923). The **mandibulo-auricularis** is usually not present as a distinct, fleshy muscle in chimpanzees; it possibly corresponds to part or to all of the stylomandibular ligament, as is usually the case in modern humans. Seiler (1974a, 1976) stated that the 'auricularis inferior' (mandibulo-auricularis *sensu* the present study) is commonly found in *P. troglodytes*, running from the fascia glandulae parotis to the region of the ear. However, the structure described and illustrated by this author in this species seems to be mainly a vestigial/ligamentous structure, and not the distinct, fleshy, mandibulo-auricularis muscle that is usually present in strepsirrhines and other mammals.

Intrinsic facial muscles of ear (Fig. 5)
- Usual attachments: See notes below.
- Usual innervation: Branches of CN7.
- Notes: The intrinsic facial muscles of the ear of chimpanzees have been rarely described in detail in the literature, and they were difficult to analyze in our specimens, but some authors such as Hubert (e.g., 1930b, 1931), Ruge (e.g., 1987ab) and Seiler (e.g., 1974a, 1976) have examined these muscles in some detail. According to Seiler (1976) the **helicis major** (part of 'depressor helicis' *sensu* Ruge 1887ab), **helicis minor**, **antitragicus**, **tragicus** (e.g., Fig. 5; 'tragus' *sensu* Gratiolet & Alix 1866 and Swindler & Wood 1973), **obliquus auriculae** (part of 'musculus auriculae proprius posterior' *sensu* Ruge 1887ab and of 'transversi et obliqui' *sensu* Huber 1930b, 1931) and **transversus auriculae** (part of 'musculus auriculae proprius posterior' *sensu* Ruge 1887ab and of 'transversi et obliqui' *sensu* Huber 1930b, 1931) are usually present in chimpanzees, as is normally the case in modern humans. He suggested that the **incisurae terminalis ('incisurae Santorini')** and the **'intercartilagineus'** are usually absent in chimpanzees, but that the **pyramidalis auriculae ('trago-helicinus'** *sensu* Edgeworth 1935 and Seiler 1974a) is usually present in these primates. He stated that the **depressor helicis** is inconstant in chimpanzees. Edgeworth (1935) stated that the depressor helicis is usually missing in chimpanzees; Ruge (1887ab) reported a 'depressor helicis' in these primates, but the structure reported by him corresponds to the helicis major and helicis minor *sensu* Seiler (1974a, 1976).

Risorius (Figs. 1–2)
- Usual attachments: From the platysma myoides to the angle of the mouth.
- Usual innervation: Branches of CN7.

- Notes: The risorius is often, but not always, present as a distinct muscle in chimpanzees. It was found in at least one side of the body of five of the eight chimpanzees dissected by us, and was reported and/or shown in other chimpanzees by various authors (e.g., illustrations of Virchow 1915, Sonntag 1923, 1924, Burrows et al. 2006, see also descriptions of Hartmann 1886, Huber 1930b, 1931, Seiler 1976).

Zygomaticus major (Figs. 1–4)
- Usual attachments: From the temporalis fascia and the zygomatic arch/bone (not from the ear, although it often originates nearer to the ear than is usually the case in modern humans) to the corner of the mouth, passing superficial to the levator anguli oris facialis.
- Usual innervation: Buccal branches of CN7 (Miller 1952: *P. paniscus*).
- Function: Moves the lips to expose the canine, premolar and molar teeth (Gratiolet & Alix 1866); elevates the lip corners superiorly and draws them laterally, increasing the angle of the mouth (Waller et al. 2006).
- Notes: Sonntag 1923 reported a 'zygomaticus major' running from the anterior end of the zygoma to the angle of the mouth and a 'zygomaticus minor' running from the zygomatic bone and the temporalis fascia to the angle of the mouth in *P. troglodytes*. This corresponds to the posterior and anterior heads of the zygomaticus major, respectively, of some of the chimpanzees dissected by us. The structure that Sonntag (1923) designated as a 'strong bundle from the orbital part of the orbicularis palpebrarum', running from the orbicularis oculi to the angle of the mouth, corresponds to the zygomaticus minor *sensu* the present study. Pellatt (1979) reported a *P. troglodytes* specimen with a 'zygomaticus major' and also a 'malaris' and a 'portion of the orbicularis oculi' that run from the orbital region to the angle of the mouth; the 'portion of the orbicularis oculi' corresponds to part of the zygomaticus minor of modern humans, while the 'malaris' might either correspond to a bundle of the zygomaticus major (in such cases the zygomaticus major would have two bundles), or to part of the zygomaticus minor. The former hypothesis (i.e., that the 'malaris' is part of the zygomaticus major) was supported by Burrows et al. (2006), who reported a zygomaticus major with a deep head attached caudally to the zygomatic arch and a superficial head attached via the skin over the superolateral portion of the face; the deep head fibres are arranged more transversely whereas the fibres of the superficial head are more oblique. In these cases the heads fuse approximately half of the way through their courses and attach together into the lateral-most portion of the orbicularis oris muscle at the modiolus, with a small attachment into the corresponding skin. In at least some of the chimpanzees dissected by us (e.g., HU PT1, PFA 1077 and PFA UNC) the zygomaticus major is similar to that described by Burrows et al. (2006) in that it is divided into a more anterior, oblique and superficial portion that attaches to the skin over the

zygomatic arch (just posteriorly to the attachment of the zygomaticus minor) and a more posterior, horizontal and deep portion that attaches to the bony zygomatic arch (its posterior region being relatively distant to the anterior region of the ear). The **depressor tarsi ('preorbicularis')** is usually not present as a distinct muscle in chimpanzees.
- Synonymy: Grand zygomatique (Gratiolet & Alix 1866); part of zygomatic mass (Macalister 1871); part or totality of zygomatici (Champneys 1872); zygomaticus plus part or totality of orbito-labialis (Sullivan & Osgood 1925); zygomaticus major and probably also zygomaticus minor (Sonntag 1923); zygomaticus (Miller 1952); zygomaticus inferior (Seiler 1976); zygomaticus major and probably also part or totality of malaris (Pellatt 1979).

Zygomaticus minor (Figs. 1, 3–4)
- Usual attachments: Mainly from the zygomatic bone and the orbicularis oculi (relatively far from the ear and near to the eye, as is usually the case in modern humans: see, e.g., plate 26 of Netter 2006) to the corner of the mouth and to the upper lip, being mainly superficial to the levator anguli oris facialis.
- Usual innervation: Buccal branches of CN7 (Miller 1952: *P. paniscus*).
- Notes:
- Synonymy: Petit zygomatique (Gratiolet & Alix 1866); part of zygomatic mass (Macalister 1871); bundle from orbital part of orbicularis palpebrarum (Sonntag 1923); zygomatic head of quadratus labii superioris (Miller 1952); zygomaticus superior (Seiler 1976); part of orbicularis oculi, and possible part or totality of malaris (Pellatt 1979).

Frontalis (Fig. 1, 3)
- Usual attachments: From the galea aponeurotica to the skin of the eyebrow and nose.
- Usual innervation: Temporal branches of CN7 (Miller 1952: *P. paniscus*).
- Function: Waller et al. (2006) reported a 'pars medialis' and a 'pars lateralis', but, as they recognized, this is a functional and not a morphological distinction (i.e., the former elevates the medial portion of the eyebrow while the latter elevates the lateral and mid portion of the eyebrow).
- Synonymy: Part of occipito-frontalis (Owen 1830-1831, Macalister 1871, Sutton 1883); probably corresponds to part of frontalis *sensu* Pellatt (1979).

Auriculo-orbitalis (Figs. 1, 5)
- Usual attachments: From the anterior portion of ear to the region of the frontalis.
- Usual innervation: Temporal branches of CN7 (Miller 1952: *P. paniscus*).
- Notes: In Terminologia Anatomica (1998) the **temporoparietalis** is considered to be a muscle that is usually present in modern humans (originating mainly from the lateral part of the galea aponeurotica and passing inferiorly to insert

onto the cartilage of the auricle, in an aponeurosis shared with the other auricular muscles). However, according to authors such as Loth (1931) the temporoparietalis is usually absent as a distinct muscle in modern humans. According to Diogo et al. (2008, 2009b), the temporoparietalis and **auricularis anterior** derive from the auriculo-orbitalis so when the temporoparietalis is not present as a distinct muscle these authors use the name auriculo-orbitalis to designate the structure that is often designated in the literature as 'auricularis anterior': that is, one can only use this latter name when the temporoparietalis is present. In various primates reported by Seiler (1976) including hominoids such as *H. moloch* and *Pongo pygmaeus*, he reported both a 'pars orbito-temporalis of the frontalis' and an 'auricularis anterior' attaching posteriorly onto the ear. That is, the structure that he designated as 'auricularis anterior' is differentiated from the auriculo-orbitalis, as is the case in various other primates, but according to him, and contrary to the condition in *Pan troglodytes* and *Gorilla gorilla* (e.g., Fig. 143 of Seiler 1976), in hominoids such as *H. moloch* and *P. pygmaeus* there is 'still' a connection between the main body of the auriculo-orbitalis and the ear. One would think that the temporoparietalis of taxa such as modern humans would probably correspond to those remaining fibers of the auriculo-orbitalis that did not differentiate into the 'auricularis anterior', and this was what Jouffroy & Saban (1971) suggested in their study on the facial muscles of mammals. However, in at least some, if not all, taxa this is clearly not the case. For example, in Fig. 143 of Seiler (1976) the structure that he designates as 'pot', which corresponds to the remaining fibers of the auriculo-orbitalis that do not form an 'auricularis anterior', does not correspond to the structure that is usually designated as temporoparietalis in modern human anatomical atlases, which as its name indicates, usually runs mainly superoinferiorly from the parietal bone to the temporal region. Therefore, in order to be as consistent as possible with our previous studies and because the 'pars orbito-temporalis of the frontalis' and the 'auricularis anterior' *sensu* Seiler (1976) are very likely derived from the same anlage and are often related to each other, being often even continuous, we consider these two structures as parts/bundles of the auriculo-orbitalis *sensu* the present study (see, e.g., Fig. 74 of Seiler 1976). That is, the 'auricularis anterior' *sensu* Seiler 1976 is considered to be part of the auriculo-labialis *sensu* the present study, except in those few primates that have a distinct temporoparietalis (in those few primates the 'auricularis anterior' *sensu* Seiler 1976 is named by us as auricularis anterior, to clearly indicate that those primates have a distinct temporoparietalis). It is however possible, and in our opinion likely, that as suggested by Jouffroy & Saban (1971) the temporoparietalis of modern humans corresponds to the 'pars orbito-temporalis of the frontalis' *sensu* Seiler (1976) and that most of the confusion related to this issue is due to erroneous descriptions of the modern human temporoparietalis in anatomical atlases (which suggest that this is

mainly a vertical muscle running superoinferiorly from the parietal bone to the temporal region, a description that does not match with the usual configuration of Seiler's 'pars orbito-temporalis of the frontalis' in other primates, which mainly runs horizontally to the region of the orbit to the region of the ear). If this is the case, then most primates would have a distinct temporoparietalis and a distinct auricularis anterior because both a 'pars orbito-temporalis of the frontalis' and an 'auricularis anterior' were reported in most primates by Seiler (1976). If this is so, then the 'pars orbito-temporalis of the frontalis' and the 'auricularis anterior' of those primates should be designated as temporoparietalis and as auricularis anterior, respectively; we plan to return to this this subject in the future. The right side of the PFA UNC specimen had a thin and broad muscle that was just medial (deep) to the auricularis superior and that connected the superior portion of the ear to the galea aponeurotica, extending superiorly to the superior attachment of the auricularis superior onto this latter aponeurosis. This thin and broad muscle seems to correspond to the temporoparietalis of modern humans (i.e., in this case there was a temporoparietalis, an auricularis superior *and* an auricularis anterior). However, on the left side of the PFA UNC specimen, and on both sides of the other chimpanzees dissected by us, we only found an auricularis superior and a single, broad muscle that extends anteriorly to, and is blended with, the frontalis. This configuration seems to us to correspond to the auriculo-orbitalis of other chimpanzees and other primates.
- Synonymy: Probably corresponds to the auricularis anterior *sensu* Gratiolet & Alix (1866), Virchow (1915), Sonntag (1923, 1924), Miller (1952), Swindler & Wood (1973), Seiler (1974a), Gibbs (1999), Burrows et al. (2006) and Burrows (2008) and to the attrahens aurem *sensu* Wilder (1862) and Macalister (1871); might correspond to part of the attolens/auricularis anterior *sensu* Sonntag (1923); probably corresponds to part of frontalis *sensu* Pellatt (1979). The **zygomatico-auricularis** is not present as a distinct muscle.

Auricularis superior (Fig. 5)
- Usual attachments: From the superior margin of the ear to the galea aponeurotica.
- Usual innervation: Temporal branches of CN7 (Miller 1952: *P. paniscus*).
- Synonymy: Atollens aurem (Wilder 1862, Macalister 1871); temporoparietalis (Virchow 1915); part or totality of attolens aurem (Sonntag 1923); part or, more likely, totality of auricularis superior et anterior (Pellatt 1979).

Orbicularis oculi (Figs. 1, 3–5)
- Usual attachments: From a continuous bony attachment around the orbit to skin near the eye. It is usually divided into a pars palpebralis and a pars orbitalis, as in modern humans.
- Usual innervation: Branches of CN7.

- Function: Inferior portion of pars orbitalis elevates the infraorbital triangle (or equivalent area) superiorly and medialwards; inferior portion lowers the mid and lateral portion of the eyebrows (but perhaps this is also done by the depressor supercilli instead; Waller et al. 2006).
- Notes: The orbicularis oculi of chimpanzees is often differentiated into a pars orbitalis and a pars palpebralis, which is in turn subdivided into a pars profunda ('pars lacrimalis') and a fasciculus ciliaris (according to Sonntag 1924 the fasciculus ciliaris is not differentiated in a few chimpanzees). Seiler (1971d, 1976) states that, within the Catarrhini, an independent muscle **'infraorbitalis'** is present in *Macaca*, *Pongo*, *Pan* and *Homo*; however, this structure seems to correspond to part of the orbicularis oculi and/or of the levator labii superioris alaeque nasi *sensu* the present study.
- Synonymy: Orbiculaire plus palpébral (Gratiolet & Alix 1866); orbicularis palpebrarum and tensor tarsi (Macalister 1871), the latter structure corresponding to the pars profunda, or lacrimalis, sensu the present study; orbicularis palpebrarum (Sutton 1883). The **zygomatico-orbicularis** is missing in chimpanzees.

Depressor supercilii (Figs. 1, 3, 5)
- Usual attachments: From the ligamentum palpebrale mediale to the eyebrow.
- Usual innervation: Branches of CN7.
- Notes: Authors such as Gratiolet & Alix (1866), Macalister (1871), Sonntag (1923), Miller (1952) and Pellatt (1979) did not describe a depressor supercilii in the *P. troglodytes* and *P. paniscus* specimens dissected by them, but as noted by Seiler (1986) and Burrows et al. (2006) and corroborated by our dissections, chimpanzees do usually have this muscle.

Corrugator supercilii (Figs. 4–5)
- Usual attachments: From the medial angle of the supraorbital ridge and the frontal process of the maxilla to the deep surface of the skin of the eyebrow region.
- Usual innervation: Temporal branches of CN7 (Miller 1952: *P. paniscus*).
- Synonymy: Sourcilier (Gratiolet & Alix 1866).

Levator labii superioris (Figs. 1, 3)
- Usual attachments: From the infraorbital region mainly to the upper lip.
- Usual innervation: Buccal branches of CN7 (Miller 1952: *P. paniscus*).
- Function: Elevates the upper lip (Waller et al. 2006).
- Synonymy: Releveur propre de la lèvre supérieure (Gratiolet & Alix 1866); part of levator labii superioris (Macalister 1871); part or totality of maxillo-labialis (Edgeworth 1935); infraorbital head of quadratus labii superioris (Miller 1952).

Levator labii superioris alaeque nasi (Figs. 1, 3)
- Usual attachments: From the region of the ligamentum infraorbitale mediale, including the lacrimal bone (e.g., Burrows et al. 2006), and/or the midline of the nasal dorsum (e.g., Virchow 1915), to the upper lip and sometimes to the ala of the nose (e.g., Gratiolet & Alix 1866, Burrows et al. 2006).
- Usual innervation: Buccal branches of CN7 (Miller 1952: *P. paniscus*).
- Function: Wrinkles the skin surrounding the nose (Waller et al. 2006).
- Notes: Virchow (1915) and Gibbs (1999) stated that in modern humans the levator labii superioris alaeque nasi originates mainly from the frontal process of the maxilla around the medial angle of the orbital opening, while in *Pan troglodytes* it takes origin from the midline of the nasal dorsum. However, as reported by authors such as Gratiolet & Alix (1966) in chimpanzees the muscle often also, or exclusively, originates from the inner angle of the eye, the frontal process of the maxilla and/or the lacrimal bone near the ligamentum infraorbitale mediale.
- Synonymy: Releveur commun de l'aile du nez et de la lèvre supérieure (Gratiolet & Alix 1866); part of levator labii superioris (Macalister 1871); angular head of quadratus labii superioris (Miller 1952).

Procerus (Fig. 5)
- Usual attachments: From the frontalis to the dorsomedial region of the nose.
- Usual innervation: Buccal branches of CN7 (Miller 1952: *P. paniscus*).
- Function: Depresses the medial corners of the eyebrows (Waller et al. 2006).
- Notes: According to Seiler (1971c, 1976) the **'depressor glabellae'** is usually present as a distinct muscle in chimpanzees, whereas the procerus is inconstant. However, the 'depressor glabellae' is often considered in the literature to be part of the procerus (see, e.g., Terminologia Anatomica 1998); in the chimpanzees dissected by us and by authors such as Gratiolet & Alix (1866), Pellatt (1979), Miller (1952) and Burrows et al. (2006) procerus was present.
- Synonymy: Pyramidalis nasi (Gratiolet & Alix 1866, Sutton 1883, Sonntag 1923); depressor glabellae (Virchow 1915); naso-labialis superficialis, pyramidalis narium, frontalis pars per dorsum nasi ducta, dorsalis narium, retractor naso-labialis or levator naso-labialis vestibularis (Jouffroy & Saban 1971); procerus plus part or totality of depressor glabellae (Seiler 1971c, 1976).

Buccinatorius (Figs. 3, 8–9, 12–13)
- Usual attachments: From the pterygomandibular raphe, the infero-lateral surface of the maxilla, and the supero-lateral border of the mandible, mainly to the angle of the mouth and the upper and lower lips.
- Usual innervation: Buccal branches of CN7 (Miller 1952: *P. paniscus*).

Nasalis (Fig. 1)
- Usual attachments: From the maxilla, deep to the orbicularis oris, to the lateral margin of the nose (pars transversa) and the lateral portion of the inferior margin of the ala of the nose (pars alaris).
- Usual innervation: Buccal branches of CN7 (Miller 1952: *P. paniscus*).
- Notes: Seiler (1970, 1971c, 1976) describes a 'nasalis' and a **'subnasalis'** in various catarrhines. The 'subnasalis' and 'nasalis' *sensu* Seiler could correspond to the pars alaris and pars transversa of the nasalis of modern human anatomy, respectively (compare, e.g., Fig. 141 of Seiler 1976 to plate 26 of Netter 2006). However, according to Seiler the 'subnasalis' while often present in chimpanzees is usually missing in *Homo* and *Gorilla*. This indicates that the subnasalis does not correspond to the pars alaris of the nasalis *sensu* the present study, because this latter structure *is* usually found in modern humans. Be that as it may, as other authors do not refer to a 'subnasalis' muscle and as we also did not find a distinct separate 'subnasalis' in the chimpanzees dissected by us, this structure probably corresponds to part of the nasalis and/or of the orbicularis oris *sensu* the present study. Seiler (1970, 1971c, 1976) also describes a **'depressor septi nasi'** and a **'musculus nasalis impar'** in a few catarrhines, but not in chimpanzees. One hypothesis is that the 'nasalis impar' corresponds to part or to the totality of the depressor septi nasi that is illustrated in a few atlases of modern human anatomy as a vertical muscle that lies on the midline and that attaches mainly onto the inferomesial margin of the nose (compare Fig. 145 of Seiler 1976 with plate 26 of Netter 2006). However, most atlases of modern human anatomy show two depressor septi nasi muscles, one in each side of the body, running obliquely (superomedially) from the upper lip to a more medial part of the inferior region of the nose. Thus, the 'nasalis impar' *sensu* Seiler (1976) corresponds to an additional midline muscle that is only inconstantly present in catarrhines, while the 'depressor septi nasi' *sensu* Seiler (1976) is effectively similar to the depressor septi nasi shown in most atlases of modern human anatomy. Table 3 of Seiler (1979b) refers to a **'labialis superior profundus'** and this suggests that according to him the 'depressor septi nasi' and the 'labialis superior profundus' are not the same structure, whereas Lightoller (1928a, 1934) suggests that these are one and the same structure (Seiler 1976 clearly shows both these structures in various primates, e.g., in his Fig. 145 of *Gorilla*, the 'labialis superior profundus' probably corresponding to part of the orbicularis oris *sensu* the present study). Jouffroy & Saban (1971) designate the nasalis as the 'naso-labialis profundus pars anterior' and the depressor septi nasi as the 'naso-labialis profundus pars mediana'; this suggests that these authors consider these two muscles derive from the same structure. In their Fig. 471, Jouffroy & Saban (1971) designate the 'labii profundus superioris' as depressor septi nasi, thus suggesting that the 'labii profundus superior' could correspond to the depressor septi nasi of modern humans as suggested by

Lightoller (1928a, 1934). However, our comparisons and dissections suggest that the homologies proposed by Seiler (1976) are somewhat doubtful and that: 1) the 'nasalis' *sensu* Seiler likely corresponds to part or the totality of the nasalis *sensu* the present study; 2) the pars transversa of the nasalis of modern humans might correspond to the 'subnasalis' *sensu* Seiler (see, e.g., Fig. 141 of Seiler 1976; however, Seiler stated that the 'subnasalis' is missing in modern humans), to the depressor septi nasi *sensu* Seiler 1976 (see, e.g., Fig. 145 of Seiler 1976), and/or to part of the nasalis *sensu* the present study (see, e.g., Fig. 145 of Seiler 1976); 3) the depressor septi nasi of modern humans corresponds to the 'depressor septi' nasi *sensu* Seiler; therefore, the 'depressor septi nasi' *sensu* Seiler does not correspond to part of the 'labialis superior profundus' *sensu* Seiler, as suggested by Lightoller (1928a, 1934; see, e.g., Fig. 141 of Seiler 1976), nor to the 'nasalis impar' *sensu* Seiler (in fact, Seiler stated that the 'nasalis impar' is missing in modern humans, which do usually have a depressor septi nasi *sensu* the present study).
- Synonymy: Myrtiformis plus transversus (Gratiolet & Alix 1866), which correspond respectively to the pars alaris and pars transversa *sensu* the present work; includes the compressor nasi *sensu* Macalister 1871, which corresponds to the pars transversa *sensu* the present work; part of nasal muscles (Sonntag 1923); part of naso-labialis (Miller 1952).

Depressor septi nasi (Fig. 3)
- Usual attachments: From the maxilla, deep to the orbicularis oris, to the inferior region of the nose.
- Usual innervation: Branches of CN7.
- Notes: Authors such as Ruge (1887a,b), Macalister (1871), Sonntag (1923), Sullivan & Osgood (1925), Miller (1952) and Pellatt (1979) did not refer to the depressor septi nasi in their descriptions of chimpanzees. However Burrows et al. (2006) did find this muscle in their careful dissections of these primates, and we also found it in at least some of the chimpanzees dissected by us (see, e.g., Fig. 3).
- Synonymy: Part of nasalis (Miller 1952).

Levator anguli oris facialis (Fig. 4)
- Usual attachments: From the canine fossa of the maxilla to the angle of mouth.
- Usual innervation: Buccal branches of CN7 (Miller 1952: *P. paniscus*).
- Notes: As proposed by Diogo et al. (2008) and Diogo & Abdala (2010), we use the name 'levator anguli oris facialis' here (and not the name 'levator anguli oris', as is usual in atlases of modern human anatomy) to distinguish this muscle from the **levator anguli oris mandibularis** (which is usually also designated as 'levator anguli oris' in the literature) found in certain reptiles, which is part of the mandibular (innervated by CN5) and not of the hyoid (innervated by

CN7) musculature. The **anomalus maxillae, anomalus nasi** and **anomalus menti**, found as anomalies in a few modern humans, are usually missing in chimpanzees.
- Synonymy: Caninus (Gratiolet & Alix 1866, Miller 1952, Burrows 2008); part of caninus *sensu* Seiler (1976), which also includes the depressor anguli oris *sensu* the present study.

Orbicularis oris (Figs. 1, 3–4)
- Usual attachments: From skin, fascia and adjacent regions of lips to skin and fascia of lips.
- Usual innervation: Buccal branches of CN7 (Miller 1952: *P. paniscus*).
- Function: Reduces the lip aperture and funnels/protrudes the lips (Waller et al. 2006).
- Notes: Huber (1931) stated that the muscles **'rectus labii inferioris'** and **'rectus labii superioris'**, although poorly differentiated, are present in *Pan*, but they are more differentiated in modern humans. He also stated that the muscles **'incisivus labii superioris'** (sometimes designated as **'pars supralabialis of the buccinatorius'**) and **'incisivus labii inferioris'** are found in modern humans, but he suggested they are not found in *Pan*. However, Seiler (1970, 1971c) describes a **'cuspidator oris'** in chimpanzees. This structure, which was designated as **'labialis superior profundus'** by Seiler 1976, probably corresponds to the **'incisivus labii superior'** *sensu* Lightoller (1928, 1934, 1939) and, thus to part of the orbicularis oris *sensu* the present study. Seiler (1976) also describes a **'labialis inferior profundus'** in chimpanzees, which thus probably corresponds to the **'incisivus labii inferioris'** *sensu* Lightoller (1928, 1934, 1939) and to part of the orbicularis oris *sensu* the present study. Lightoller (1925) stated that a differentiated pars marginalis of the orbicularis oris is present in modern humans but not in *Pan*, the presence of this pars marginalis being probably related to the evolution of speech in modern humans. Pellatt (1979) also stated that the 'peripheral' and 'marginal' parts of the orbicularis oris are not clearly differentiated in chimpanzees.
- Synonymy: Orbiculaire des lèvres (Gratiolet & Alix 1866); orbicularis oris plus incisive muscles (Sonntag 1923); orbicularis oris plus rectus labii inferioris and rectus labii superioris (Huber 1930b, 1931).

Depressor labii inferioris (Figs. 1, 3)
- Usual attachments: From the platysma myoides and the mandible to the lower lip.
- Usual innervation: Mandibular branches of CN7 (Miller 1952: *P. paniscus*).
- Function: Depresses the medial portion of the lower lip (Waller et al. 2006).
- Notes: Some authors (e.g., Gratiolet & Alix 1866, Champneys 1872) have suggested that in chimpanzees the depressor labii inferioris is not present as a distinct muscle. However, the descriptions of Hartmann (1886), Sonntag

(1923), Miller (1952), Seiler (1976), Pellatt (1979) and Burrows et al. (2006) and our own dissections show that the depressor labii inferioris is usually present as a distinct muscle in these primates. The depressor labii inferioris usually arises from the mandible and inserts onto the skin of the lower lip, being mainly blended with its counterpart as well as with the orbicularis oris, depressor anguli oris, mentalis and platysma myoides (it is mainly distinct from the platysma myoides because, in lateral view, its fibers run more obliquely than the mainly anteroposteriorly oriented anterior fibers of this latter muscle).
- Synonymy: Quadratus labii inferioris (Sonntag 1923, 1924, Miller 1952, Jouffroy & Saban 1971).

Depressor anguli oris (Figs. 1–3)
- Usual attachments: From the angle of mouth to the fascia of the platysma myoides.
- Usual innervation: Mandibular branches of CN7 (Miller 1952: *P. paniscus*).
- Function: Pulls the corner of the mouth downward (Waller et al. 2006).
- Notes: Authors such as Gratiolet & Alix (1866), Huber (1931) and Jouffroy & Saban (1971) stated that in some chimpanzees the depressor anguli oris extends inferomedially to the inferior margin of the mandible to meet its counterpart in the ventral midline, thus forming a **transversus menti** (Loth 1931 suggested that this structure if present in 18% of chimpanzees).
- Synonymy: Triangularis (Gratiolet & Alix 1866, Sonntag 1923, 1924, Edgeworth 1935, Miller 1952); part of caninus *sensu* Seiler (1976), which also includes the levator anguli oris facialis.

Mentalis (Fig. 4)
- Usual attachments: From the mandible to the skin below the lower lip, being mainly deep to the orbicularis oris and to the depressor labii inferioris, sometimes meeting with its counterpart in the midline (e.g., Gibbs 1999).
- Usual innervation: Branches of CN7.
- Function: Pushes the skin of the chin area superiorly (Waller et al. 2006).
- Synonymy: Elevator labii inferioris proprius sensu Tyson (1699).

3.3 Branchial musculature

Stylopharyngeus (Figs. 9, 11, 14)
- Usual attachments: From the styloid process and/or adjacent regions of the cranium (e.g., Dean 1984; our dissections of HU PT1, PFA 1077, PFA UNC) to the pharyngeal wall and sometimes also the greater horn of the hyoid bone and/or the thyroid cartilage (e.g., Gratiolet & Alix 1866), passing between the middle and superior pharyngeal constrictors.
- Usual innervation: Glossopharyngeal nerve (Sonntag 1923: *P. troglodytes*).
- Notes: The **ceratohyoideus** and **petropharyngeus** are not present as distinct muscles in chimpanzees.

Trapezius (376.2 g; Figs. 8–9)
- Usual attachments: Mainly from the vertebral column (e.g., in *P. troglodytes* extending posteriorly to T12 according to Gratiolet & Alix 1866 and to Schück 1913b and even 'T13' according to Sonntag 1923, but only extending to T10 according to Macalister 1871 and Swindler & Wood 1973 and to T8 according to Stewart 1936; in *P. paniscus* extending to T9 according to Miller 1952) to the spine and acromion of the scapula and to the lateral 1/3, or more than the lateral 1/3, of the clavicle (NB the ligamentum nuchae is usually not present as a distinct, well-defined structure in chimpanzees).
- Function: Pars ascendens of trapezius exhibited EMG activity during arm-raising in *Pan* (Tuttle & Basmajian 1976); EMG indicated that the cranial trapezius was not involved in arm-raising, its recruitment being instead closely tied to head position (Larson et al. 1991).
- Notes: As in modern humans, in chimpanzees the trapezius is usually differentiated into a pars descendens, a pars transversa and a pars ascendens.
- Usual innervation: XI and C3 in one chimpanzee specimen and XI, C2, C3 and C4 in another chimpanzee specimen (Schück 1913b); spinal accessory nerve and third and fourth cervical nerves (Sonntag 1923: *P. troglodytes*); spinal accessory nerve and a branch from the anterior division of the third cervical nerve through the branchial plexus (Miller 1952: *P. paniscus*); motor innervation from the spinal accessory nerve (Swindler & Wood 1973: *P. troglodytes*).
- Synonymy: Cucullaris (Tyson 1699).

Sternocleidomastoideus (172.8 g; Figs. 6–10, 19)
- Usual attachments: The caput sternomastoideum runs from the sternum to the mastoid region and the lateral portion of the superior nuchal line; the caput cleidomastoideum runs from the medial portion of the clavicle (medial 1/4 according to Gratiolet & Alix 1866, medial 1/3 according to Sonntag 1923 and to our dissections of the specimens PFA 1016, PFA 1009 and PFA 1051) to the mastoid region; in a specimen dissected by Sutton 1883 there was also an insertion onto the atlas.
- Usual innervation: XI in one chimpanzee specimen and XI and C2 in another chimpanzee specimen (Schück 1913b); spinal accessory nerve and a branch from the anterior (ventral) division of the second cervical nerve through the cervical plexus (Miller 1952: *P. paniscus*).
- Notes: In a few chimpanzees there is a structure that is sometimes designated as muscle **'cleido-occipitalis'** (e.g., Sonntag 1923, Miller 1952) but this structure is usually deeply blended with the sternocleidomastoideus, and does not form a distinct, separate, muscle, i.e., it is instead a bundle of the sternocleidomastoideus, which is designated in the present work as **caput cleido-occipitale**). According to Wood (1870) and to the recent work of Mustafa (2006) only about 36% and 33% of modern humans, respectively, have a caput cleido-

occipitale of the sternocleidomastoideus. Among the chimpanzees dissected we only found a well-defined caput 'cleido-occipitale' of the sternocleidomastoideus on the left side of the infant chimpanzee PFA 1077, running from about the medial 1/3 of the clavicle to the occipital region, passing posteriorly and deep to the sternomastoid head and superficially to both the accessory nerve (XI) and to the cleidomastoid head, which lies deep to this nerve. Superiorly the cleido-occipitale and sternomastoid heads were blended at their attachments onto the occipital region, but the cleidomastoid head is well-separated from them by fascia and by the accessory nerve. Inferiorly these three structures are somewhat blended with each other. On the right side of this PFA 1077 specimen the cleido-occipitale and sternomastoid heads are almost completely fused to each other, the structure formed by them thus being superficial to the accessory nerve and running from the clavicle and sternum to the occipital region of the skull. This indicates that the modern human sternomastoid head might corresponds to the sternomastoid plus cleido-occipitale heads of the left side of PFA 1077 (for more discussion of this issue see Diogo & Wood 2012).

Constrictor pharyngis medius (Fig. 12)
- Usual attachments: From the dorsal raphe of the pharyngeal wall to the angle between the greater and lesser horns of this bone (e.g., Gratiolet & Alix 1866, Sonntag 1923). Thus, there is a pars ceratopharyngea, but no pars chondropharyngea, unless one considers that the portion of the muscle that inserts onto the body of the hyoid bone includes the pars chondropharyngea. Himmelreich (1977) shows in his Fig. 11a *P. troglodytes* specimen with a 'hyopharyngeus' (which corresponds to part or to the totality of the middle constrictor *sensu* the present study) that mainly inserts onto the greater horn of the hyoid bone. His Fig. 10 of *Macaca mulatta* shows, instead, a constrictor pharyngis medius divided into a pars ceratopharyngea and a thin pars chondropharyngea going to the greater and lesser horns of the hyoid bone, as is often the case in various other primates including modern humans. This indicates that at least in Himmelreich's chimpanzee specimen there is no pars distinct chondropharyngea (it was not possible to discern if this latter structure was or not present in the chimpanzees dissected by us).
- Usual innervation: Data are not available.
- Synonymy: Part or totality of hyopharyngeus (Himmelreich 1977).

Constrictor pharyngis inferior (Figs. 9, 11–12, 15)
- Usual attachments: From the dorsal raphe of the pharyngeal wall to the thyroid cartilage (pars thyropharyngea), the cricoid (pars cricopharyngea) cartilage, and/or in some cases to the fist tracheal ring (e.g., Gratiolet & Alix 1866).
- Usual innervation: Both the superior laryngeal and recurrent laryngeal nerves (Kohlbrügge 1890–1892: *H. moloch, H. agilis, H. syndactylus*).

- Synonymy: Thyropharyngeus plus cricopharyngeus (Starck & Schneider 1960, Himmelreich 1977).

Cricothyroideus (Figs. 7, 9, 11–13, 15)
- Usual attachments: From the cricoid cartilage to the thyroid cartilage, specifically to its inferior and inferomedial portion as well to a small area on the lateral surface of its inferior horn; the muscle sometimes contacts its counterpart in the midline.
- Usual innervation: External branch of the superior laryngeal nerve (Jordan 1971a,b,c).
- Notes: In chimpanzees the cricothyroideus is usually divided into a more superficial and less oblique pars recta and a deeper and more oblique pars obliqua, as is the case in modern humans. According to Gratiolet & Alix (1866), Kohlbrügge (1896), Duckworth (192), Kelemen (1948, 1969), Starck & Schneider (1960), Saban (1968), Jordan (1971a,b,c) the pars interna is also present in *Pan*, whereas Harrison (1995) stated he only found a pars interna in one of the eight chimpanzees dissected by him. In at least four of the six specimens in which we analyzed this feature in detail there was also a pars interna (i.e., PFA 1016, PFA 1009, PFA 1051 and HU PT1, the other two specimens being PFA 1077 and PFA UNC). The ***thyroideus transversus*** is usually not present in chimpanzees.
- Synonymy: Cricothyreoideus anticus (Kohlbrügge 1896).

Constrictor pharyngis superior (Fig. 14)
- Usual attachments: From the pterygoid plate/hamulus (pars pterygopharyngea), the pterygomandibular ligament/raphe (pars buccopharyngea), the mandibular region near the mylohyoid line (pars mylopharyngea), and the tongue (pars glossopharyngea), and in some cases from the basiocciput, the mucous membrane of the floor of the mouth and/or the buccopharyngeal aponeurosis (e.g., Sonntag 1923, Gibbs 1999), to the dorsal raphe of the pharyngeal wall.
- Usual innervation: Data are not available.
- Notes: The **pterygopharyngeus** is usually not present as a distinct muscle in chimpanzees.
- Synonymy: Cephalopharyngeus or gnathopharyngeus (Himmelreich 1977).

Palatopharyngeus (Figs. 14–15)
- Usual attachments: This muscle has almost never been described in detail in chimpanzees. It was reported but not described in detail by Sonntag (1923), and reported by Gratiolet & Alix (1866) as a muscle running from the soft palate to the thyroid cartilage. In our specimen PFA 1016 the muscle runs mainly from the soft palate to the to the pharyngeal wall, as is usually the case in modern humans, and is seemingly divided into a main body and a **'sphincter palatopharyngeus' (palatopharyngeal sphincter,** or **Passavant's ridge)** (Fig. 14), as is often the case in modern humans (see, e.g., pl. 65 of Netter 2006).

- Usual innervation: Data are not available.
- Synonymy: Pharyngostaphylins (Gratiolet & Alix 1866).

Musculus uvulae
- Usual attachments: This muscle has almost never been described in detail in chimpanzees. It was reported by Gratiolet & Alix (1866) as a muscle running from the posterior nasal spine of the palate (as it usually does in modern humans) to the midline, finishing in a membrane without attaching to the uvula. It was also reported by Sonntag (1923) as a muscle that does not really attach directly onto the uvula. Our chimpanzee specimens seem to have a musculus uvulae, but we could not analyze its attachments in detail.
- Usual innervation: Data are not available.
- Synonymy: Azygos uvulae (Sonntag 1923, 1924); medialis veli palatini (Himmelreich 1971).

Levator veli palatini (Figs. 14–15)
- Usual attachments: This muscle was almost never described in detail in chimpanzees. Gratiolet & Alix (1866) stated that it originates from the Eustachian tube and that it is more horizontal than in modern humans, due to the prognathism of the face of *Pan*. Sonntag (1923) wrote that it originates together with the tensor veli palatini from the apex of the petrous temporal bone, the Eustachian tube and the scaphoid fossa and that it then separates from the tensor veli palatini, to run downwards and forwards and spread out between layers of the palatopharyngeus, being more horizontal than in modern humans. Dean (1985) reported an origin from the medial aspect of the Eustachian tube and the adjacent part of the petrous apex and an insertion onto the superior surface of the soft palate. In our specimens the muscle is more horizontal than in modern human adults, running from the Eustachian region and adjacent areas of the neurocranium to the soft palate, being medial to the tensor veli palatini and lateral to the palatopharyngeus, with exception to the specimens PFA 1077 and PFA UNC, in which the muscle seemed to pass partially lateral to the tensor veli palatini.
- Usual innervation: Data are not available.
- Synonymy: Péristaphylin interne (Gratiolet & Alix 1866).

Salpingopharyngeus (Figs. 14–15)
- Usual attachments: This muscle was almost never reported in chimpanzees. In our specimens PFA 1016, PFA 1009, PFA 1051 and HU PT1 the salpingopharyngeus seems to be similar to that of modern humans. In our specimen PFA 1077 we could not find a separate muscle salpingopharyngeus, but in PFA UNC there is seemingly at least a salpingopharyngeal fold such as that found in adult modern humans, although we could not discern if this fold is associated with a separate, well-defined muscle salpingopharyngeus.
- Usual innervation: Data are not available.

Thyroarytenoideus (Figs. 16–17)
- Usual attachments: See notes below.
- Usual innervation: Inferior laryngeal nerve (from recurrent laryngeal nerve) (Jordan 1971a,b,c: *P. troglodytes*).
- Notes: With respect to hominoids and other primates there has been controversy regarding the homologies of the thyroarytenoid bundles and the presence/absence of a distinct **musculus vocalis**. Kohlbrügge (1896) dissected gorillas, chimpanzees and orangutans as well as taxa such as *Cebus*, *Semnopithecus*, *Hylobates* and *Macaca*, and stated that he could not find a distinct attachment of the thyroarytenoideus onto a true vocal cord (such as that found in modern humans) in any of these taxa, except perhaps in *Pongo*; within all the taxa mentioned above, he found an attachment onto the cricoid cartilage in *Hylobates* and *Colobus*. Giacomini (1897) examined the larynx of a gorilla and of an *Hylobates lar* specimen and based on these results and on his previous studies, he stated that only in modern humans is there a distinct, well-developed, musculus vocalis directly connected to the vocal cord, although in at least some *Hylobates* and seemingly in some *Pan*, there are a few fibers of the thyroarytenoideus that are somewhat isolated and situated near the vocal cord (see, e.g., Fig. 2 of his Plate II). Duckworth (1912) examined specimens from all of the five extant hominoid genera, as well as from cadavers belonging to *Macaca*, *Cebus*, *Semnopithecus* and *Tarsius*, and suggested that a well-developed, distinct musculus vocalis associated with the plica vocalis is also only consistently present in modern humans. But Duckworth did suggest that great apes, and particularly chimpanzees, show a configuration that is somewhat similar to that found in modern humans, in that they have a poorly developed/differentiated musculus vocalis (see, e.g., his Figs. 24 and 17). According to him, in modern humans the superior portion of the thyroarytenoideus usually includes a distinct structure that is often associated with the region of the ventriculus (this corresponds to the **'musculus ventricularis laryngis'** *sensu* Kelemen 1948, 1969), although a somewhat similar configuration is also seen in gorillas and in chimpanzees. Loth (1931) argued that in non-hominoid primates the vocal cords mainly consist of a well-developed fold(s) of mucous membrane, which are not in contact with the musculus vocalis. He stated that in hominoids such as *Pan* and *Gorilla* the folds are smaller and the musculus vocalis does not connect to the folds; such a connection was found in most modern humans. Edgeworth (1935) defended the notion that a musculus vocalis is found in some non-human primates and suggested that when this structure is present the thyroarytenoideus becomes a **'thryroarytenoideus lateralis'** because its inferior/mesial part gives rise to the vocalis muscle. Starck & Schneider (1960) described a 'pars lateralis' as well as a 'pars medialis' of the musculus vocalis that usually goes to the vocal fold/cord in *Pan*, *Pongo* and *Gorilla*, but not in *Hylobates*; the latter corresponds to the musculus vocalis of modern humans. They did not

find a pars aryepiglottica or a pars thyroepiglottica of the thyroarytenoideus in *Hylobates* and *Pongo*, but stated that other authors did report at least one of these structures in *Gorilla* and *Pan*. Saban (1968) clarified the nomenclature of the thyroarytenoideus and suggested that this muscle may be divided into the following structures: 1) a pars superior (often designated as **'thyroarytenoideus superior'**, 'thyroarytenoideus lateralis' or **'ventricularis'**); 2) a pars inferior (often designated as **'thyroarytenoideus inferior'**, **'thyroarytenoideus medialis'**, or musculus vocalis); 3) a **'ceratoarytenoideus lateralis'**; 4) a 'pars intermedia' (but this name was only used by a few authors such as Starck & Schneider 1960 who stated that some primates might have a pars superior, a pars inferior, and a pars intermedia); 5) a pars thyroepiglottica; 6) a pars aryepiglottica; 7) a pars arymembranosa; and 8) a pars thyromembranosa. According to Saban (1968) the 'ceratoarytenoideus lateralis' is usually fused with (but not differentiated from, as suggested in some anatomical atlases) the cricoarytenoideus posterior, being only a distinct muscle in a few taxa and, within primates, in *Pan* (and still in this case this seems to constitute a variant/anomaly) where it is a small muscle running from the dorsal face of the inferior thyrohyoid horn to the arytenoid cartilage. Also according to Saban the pars superior and pars inferior are more superior and inferior, respectively, in apes and modern humans (in modern humans the more inferior or medial part, i.e., the pars inferior is well developed and often designated as musculus vocalis), whereas in primates such as *Macaca* and *Papio* they are more lateral/medial; in *Pongo* the pars inferior is well-developed and lies anterior to the vocal cord, but it is not associated with it. Aiello & Dean (1990) stated that in non-human hominoids the pars aryepiglottica is often reduced in size or absent. Our observations and review of the literature indicate that the pars superior and pars inferior of the thyroarytenoideus are usually present in chimpanzees. The ceratoarytenoideus lateralis is present in at least some *Pan* according to Saban (1968) and to Macalister (1871), but was not found in the chimpanzees dissected by Jordan (1971a,b,c) or in those dissected by us. The pars intermedia is not present in chimpanzees according to Starck & Schneider (1960), to Jordan (1971a,b,c), and to our dissections. The **pars thyroepiglottica** is present in at least some of the chimpanzees dissected by us (Fig. 16) and by Gratiolet & Alix (1866) and Kelemen (1948, 1969), but it was not present in the chimpanzees examined by Jordan (1971a,b,c) and Avril (1963). The **pars aryepiglottica** is seemingly absent in our specimens and is only present in one of the 10 chimpanzees reported by Avril (1963), being absent in the chimpanzees examined by Jordan (1971a,b,c), but present in the specimen of Gratiolet & Alix (1866), in a specimen of Sonntag (1923), and in the specimen of Kelemen (1948, 1969). The **pars thyromembranosa** and **pars arymembranosa** are not present in the chimpanzees dissected by Jordan (1971a,b,c) and seemingly by Avril (1963).

- Synonymy: Thyroarytenoideus plus musculus vocalis (Duckworth 1912, Jordan 1971a,b,c).

Musculus vocalis (see thyroarytenoideus above)
- Usual attachments: See thyroarytenoideus above.
- Usual innervation: See thyroarytenoideus above.
- Notes: See thyroarytenoideus above.

Cricoarytenoideus lateralis (Figs. 16–17)
- Usual attachments: From the anterior portion of the cricoid cartilage to the arytenoid cartilage.
- Usual innervation: Inferior laryngeal nerve (from recurrent laryngeal nerve) (Jordan 1971a,b,c: *P. troglodytes*).

Arytenoideus transversus (Figs. 16–17)
- Usual attachments: Unpaired, transversly-oriented muscle connecting the two contralateral arytenoid cartilages.
- Usual innervation: Inferior laryngeal nerve (from recurrent laryngeal nerve) (Jordan 1971a,b,c: *P. troglodytes*).

Arytenoideus obliquus (Figs. 16–17)
- Usual attachments: From the arytenoid cartilage to the contralteral arytenoid cartilage; its fibers are more oblique than those of the arytenoideus transversus.
- Usual innervation: Inferior laryngeal nerve (from recurrent laryngeal nerve) (Jordan 1971a,b,c: *P. troglodytes*).
- Notes: The arytenoideus obliquus was found in *Pan* by Sonntag (1923, 1924), Gratiolet & Alix (1966), and Avril (1963) and it was present in some of the chimpanzees dissected by us; it was not reported in the specimens of Kohlbrügge (1896), Kelemen (1948), Körner (1884), Starck & Schneider (1960), and Jordan (1971a,b,c).

Cricoarytenoideus posterior (Figs. 16–17)
- Usual attachments: From dorsal portion of the cricoid cartilage to the arytenoid cartilage. There is usually no attachment onto the inferior horn of the thyroid cartilage; thus there is usually no **ceratocricoideus** *sensu* Harrison (1995) in chimpanzees.
- Usual innervation: Inferior laryngeal nerve (from recurrent laryngeal nerve) (Jordan 1971a,b,c: *P. troglodytes*).
- Notes: In 8 of the 10 chimpanzees from which we could retrieve detailed information (i.e., one specimen shown in plate 57 of Swindler & Wood 1973; one specimen shown in Fig. 2 of plate 9 of Gratiolet & Alix 1866; one specimen reported by Sonntag 1923; one specimen described by Kelemen 1948; and the six specimens dissected by us where we studied this feature in detail), there was no contact between the cricoarytenoideus posteror and its counterpart in the midline. The two exceptions are two infants dissected by us (PFA 1077,

PFA UNC), in which a few fibers ran between the two muscles. It should also be noted that Jordan 1971a,b,c examined 10 chimpanzees and stated that in a few specimens the cricoarytenoideus posterior *did* meet its counterpart in the dorsal midline, but he did not specified the number of specimens in which this occurred.
- Synonymy: Cricoarytenoideus posticus (Kohlbrügge 1896).

3.4 Hypobranchial musculature

Geniohyoideus (Figs. 11–13)
- Usual attachments: From the mandible to the body, and usually also part of the greater horn, of the hyoid bone (e.g., Gratiolet & Alix 1866, Sonntag 1923), lying close to and often merging with its counterpart at the midline (e.g., Duvernoy 1855-1856, Sonntag 1923, Edgeworth 1935 and also seen in the two chimpanzees in which we investigated this feature in detail, i.e., in PFA 1077 and PFA UNC).
- Usual innervation: Hypoglossal nerve (Sonntag 1923: *P. troglodytes*); ventral ramus of the first cervical nerve through the hypoglossal nerve (Miller 1952: *P. paniscus*).

Genioglossus (Figs. 11–12)
- Usual attachments: From the mandible to the tongue and also to the hyoid bone (e.g., our dissections and also descriptions of Sonntag 1923 and Edgeworth 1935, who referred to the fibers that attach onto the hyoid bone as part of a **'genio-hyoglossus'**). The genioglossus muscles are usually well-separated by a midline fibrous septum and/or fatty tissue (e.g., Sonntag 1923; our observations).
- Usual innervation: Hypoglossal nerve (Sonntag 1923: *P. troglodytes*).
- Notes: The muscles **genio-epiglotticus, glosso-epiglotticus, hyo-epiglotticus,** and **genio-hyo-epiglotticus,** described by authors such as Edgeworth (1935) and Saban (1968) in some primate and non-primate mammals, do not seem to be present as distinct structures in the chimpanzees dissected by us (according to these authors these muscles are usually not present in catarrhine taxa).

Intrinsic muscles of tongue
- Usual attachments: To our knowledge, there are no detailed published descriptions of these muscles in chimpanzees and we could not examine them in detail in our dissections. However, the **longitudinalis superior, longitudinalis inferior, transversus linguae** and **verticalis linguae** are consistently found in modern humans and at least some other primate and non-primate mammals, so these four muscles are very likely also present in chimpanzees. Detailed studies of the tongue and its muscles in chimpanzees are clearly needed.
- Usual innervation: Data are not available.

Hyoglossus (Figs. 7–8, 11–12, 14)
- Usual attachments: The hyoglossus is usually differentiated into a **ceratoglossus** and a **chondroglossus** (our dissections; see, e.g., Figs. 11, 14) as is usually the case in modern humans (see, e.g., Terminologia Anatomica 1998). The ceratoglossus connects the greater horn of the hyoid bone to the tongue, while the chondroglossus mainly connects the body of the hyoid bone (the inferior horn is usually poorly developed or absent) to the tongue. Two bundles of the hyoglossus have almost never been described in *Pan*, although in a few cases the hyoglossus is blended with the thyrohyoideus (e.g., Sonntag 1923).
- Usual innervation: Data are not available.

Styloglossus (Figs. 7–10, 12)
- Usual attachments: From stylohyoid process and/or from adjacent regions of the skull (e.g., Dean 1984).
- Usual innervation: Data are not available.
- Notes: As explained by Gibbs (1999) and confirmed by our dissections, contrary to the vast majority of eutherian mammals (e.g., Fig. 369 of Saban 1968) and to most primates, including other hominoids, in chimpanzees and also in modern humans (e.g., plate 59 of Netter 2006) the styloglossus runs mainly longitudinally to insert onto the tongue, but it also has a distinct oblique slip that runs anteroinferiorly at about 45° from the main body of the muscle to insert more inferiorly onto the lateral surface of the hyoglossus.

Palatoglossus (Figs. 14–15)
- Usual attachments: From the posterior portion of tongue to the soft palate.
- Usual innervation: Data are not available.
- Notes: As noted by Duchin (1990) and Gibbs (1999) and confirmed by our dissections, in chimpanzees the palatoglossus mainly attaches to the posterior portion of the tongue, whereas in modern humans it attaches to both the posterior portion and the lateral surface of the tongue.
- Synonymy: Glossopalatinus (Gratiolet & Alix 1866, Himmelreich 1971).

Sternohyoideus (LSB 9.8 g; Figs. 6–8, 18–19)
- Usual attachments: From the sternum and adjacent regions to the hyoid bone; the muscle usually does not contact or lie just next to its counterpart for most of its length (Gratiolet & Alix 1866, Sonntag 1923, Miller 1952, Starck & Schneider 1960, Swindler & Wood 1973; see, e.g., Fig. 30 of Sonntag 1923 and plate 51 of Swindler & Wood 1973).
- Usual innervation: Branches from the ansa hypoglossi composed of the anterior divisions of the first three cervical nerves (Miller 1952: *P. paniscus*).
- Notes: In chimpanzees the sternohyoideus often has transverse tendinous intersections (e.g., Gratiolet & Alix 1866, Champneys 1872, Starck & Schneider 1960) but no intersections were found in the specimens dissected by us.

Omohyoideus (LSB 3.6 g; Figs. 6–8, 10, 18)

- Usual attachments: From the scapula to the hyoid bone, passing deep to the sternocleidomastoideus.
- Usual innervation: Branches from the ansa hypoglossi composed of the anterior divisions of the first three cervical nerves (Miller 1952: *P. paniscus*).
- Notes: As in modern humans, in chimpanzees the omohyoideus has a well-defined intermediate tendon between the superior head and the inferior head/heads (e.g., Vrolik 1841, Gratiolet & Alix 1866, Bischoff 1870, Macalister 1871, Sonntag 1923, 1924, Ashton & Oxnard 1963; our dissections). Contrary to the situation in most other primates, in at least some specimens of *Gorilla*, *Pan* and *Homo* the omohyoideus has three bellies (usually a superior belly, an inferomedial belly, and an inferolateral belly: see Figs. 7 and 8 and Diogo & Wood 2011, 2012). This condition was reported in chimpanzees by Gratiolet & Alix (1866), who designated one of the inferior bellies as 'cleidohyoideus', by Sonntag (1923; see, e.g., his Fig. 32), and it was found in two out of the chimpanzees dissected by us (PFA 1016 and PFA 1009).
- Synonymy: Coracohyoideus (Tyson 1699), and the omohyoideus probably corresponds to the omo-hyoïdien plus cléido-hyoïdien *sensu* Gratiolet & Alix (1866).

Sternothyroideus (LSB 2.1 g; Figs. 7, 9–10, 13, 18)
- Usual attachments: From the sternum and adjacent regions (e.g., rib 1: Gratiolet & Alix 1866, Sonntag 1923) to the thyroid cartilage; the muscle do not usually contact their counterparts at the midline.
- Usual innervation: Branches from the ansa hypoglossi composed of the anterior divisions of the first three cervical nerves (Miller 1952: *P. paniscus*).
- Notes: In chimpanzees, the sternothyroideus may have transverse tendinous intersections (e.g., Macalister 1871), but most specimens do not have such intersections (Champneys 1872; our dissections). Contrary to the condition in modern humans and in most other primates, in chimpanzees the main body of the sternothyroideus usually extends anteriorly to the posterior portion of the main body of the thyrohyoideus (e.g., Gratiolet & Alix 1866, Sonntag 1923, Starck & Schneider 1960, Swindler & Wood 1973; our dissections).

Thyrohyoideus (LSB 1.3 g; Figs. 9–11, 13, 15, 18)
- Usual attachments: From the thyroid cartilage to the hyoid bone.
- Usual innervation: Branch of the hypoglossal nerve (Sonntag 1923: *P. troglodytes*); branch of the anterior division of the first cervical nerve through the descendens hypoglossi (i.e., branch of the hypoglossal nerve) (Miller 1952: *P. paniscus*).
- Notes: Swindler & Wood (1973) stated that in *Pan* there is no pyramidal lobe of the thyroid gland and we and others found no evidece of a **levator glandulae thyroideae** muscle, which is present as an anomaly in modern humans. However, ratiolet & Alix (1866) described a *P. troglodytes* specimen in which the

cricothyroideus has three divisions: a pars recta, a pars obliqua, and a 'small additional bundle' (see their plate 8) running superficially to these divisions, from the inferior margin of the thyroid cartilage to the more anterior tracheal cartilages. In view of its position and of being superficial to the cricothryroideus, this 'additional bundle' could correspond to the levator glandulae thyroideae of modern humans, but this latter muscle more often originates from the hyoid bone than from the inferior margin of the thyroid cartilage.

3.5 Extra-ocular musculature

Muscles of eye
- Usual attachments: To our knowledge, there are no detailed published descriptions of these muscles in chimpanzees. The **rectus inferior, rectus superior, rectus medialis, rectus lateralis, obliquus superior, obliquus inferior, orbitalis** and **levator palpebrae superioris** are consistently found in modern humans and at least some other primate and non-primate mammals, so these muscles are very likely also present in chimpanzees. Detailed studies of the eye and the extraocular muscles in chimpanzees and other apes are clearly needed.
- Usual innervation: Data are not available.

CHAPTER 3

Pectoral and Upper Limb Musculature

Serratus anterior (LSB 359.3 g; Figs. 19, 25)
- Usual attachments: From the medial side of the scapula, being well-separated from the levator scapulae, to the ribs (in *P. troglodytes* to, e.g., ribs 1, 3 and 4–11: Wilder 1862; ribs 1–10: Gratiolet & Alix 1866, Champneys 1872; ribs 1–12: Macalister 1871; ribs 1–11: Hepburn 1892, Sonntag 1923; ribs 1–10 or 1–12: Schück 1913b; 1–11 or 1–12: Stewart 1936; extending to rib 12: Stern et al. 1980b; in *P. paniscus* to ribs 1–10: Miller 1952).
- Usual innervation: Long thoracic nerve, from C4, C5 and C6 (Hepburn 1892: *P. troglodytes*); long thoracic nerve, from C6, C6 and C7 (Schück 1913b: *P. troglodytes*); long thoracic nerve from the brachial plexus (Miller 1952: *P. paniscus*; Swindler & Wood 1973: *P. troglodytes*).
- Function: Caudal part of muscle exhibited EMG activity during arm-raising (Tuttle & Basmajian 1976: *P. troglodytes*); EMG activity mainly related to arm-raising motions, except lowest parts of muscle, which display significant activity during suspensory postures and locomotion presumably to control the tendency of the scapula to shift cranially relative to the rib cage (Larson et al. 1991: *P. troglodytes*).
- Synonymy: Serratus magnus (Wilder 1862, Macalister 1871, Champneys 1872, Hepburn 1892, Beddard 1893, Sonntag 1923, 1924); grand dentelé (Gratiolet & Alix 1866, Broca 1869); pars caudalis of serratus anticus (Schück 1913b); caudal serratus anterior (Larson et al. 1991).

Rhomboideus (LSB 99.9 g; Fig. 26)
- Usual attachments: From the medial border of the scapula to the cervical and thoracic vertebrae (in *P. troglodytes* to, e.g., C3–T8 or C3–T7: Schück 1913b; C6–T4: Sonntag 1923, 1924; C2–T6: Stewart 1936; C4–T6: Swindler & Wood 1973; often from C2 or C3 to T6 or T7: Andrews & Groves 1976; C4–T3, T4 or T5 in our specimen PFA 1077; C6–T5 in our specimen PFA UNC; in *P. paniscus* to C3-T6: Miller 1952).
- Usual innervation: C4 (Hepburn 1892: *P. troglodytes*); C4 and C5 (Champneys 1872: *P. troglodytes*; Sonntag 1923: *P. troglodytes*); C5 and dorsal scapular nerve

(Miller 1952: *P. paniscus*); C3 and C4 (in one specimen) or C4 and C5 (in one specimen) (Schück 1913b: *P. troglodytes*); dorsal scapular nerve (Swindler & Wood 1973: *P. troglodytes*).
- Function: The rhomboideus exhibited EMG activity during arm-raising, but there was little or no activity during hoisting and suspension (Tuttle & Basmajian 1978a: *P. troglodytes*).
- Notes: The **rhomboideus major, rhomboideus minor** and **rhomboideus occipitalis** are usually not present as distinct entities in chimpanzees; the rhomboideus of chimpanzees probably corresponds to the rhomboideus major plus rhomboideus minor of modern humans, the rhomboideus occipitalis being missing in both modern humans and chimpanzees (e.g., Diogo & Wood 2011, 2012). However, some exceptions are worthy of notice. Gratiolet & Alix (1866) reported a *P. troglodytes* specimen with an undivided rhomboideus originating from the first 7 vertebral spines, the 'cervical ligament', the axis and the occipital region, but the occipital origin was only from an aponeurosis, so there was no distinct rhomboideus occipitais. Champneys (1872) described a rhomboideus major (innervated by C5) and a rhomboideus minor (innervated by C4) in the *P. troglodytes* specimen dissected by him; according to him the rhomboideus major is similar to that of modern humans, its tendon of origin being fused with the overlying rhomboideus minor. Ziegler (1964) referred to, and showed in his Fig. 4, a 'rhomboideus minor' and a 'rhomboideus major' in the *P. troglodytes* specimen dissected by him, but he did not describe these structures in detail so it cannot be discerned if they are well-defined and separated and thus if they effectively correspond to the rhomboideus major and rhomboideus minor of modern humans.

Levator scapulae (LSB 52.8 g; Figs. 12, 23)
- Usual attachments: From the cervical vertebrae to the anterior portion of the medial side of the scapula.
- Usual innervation: C4 (Champneys 1872: *P. troglodytes*); one specimen by C4 while in other it was by C3 and C4 (Schück 1913b: *P. troglodytes*); C3 and C5 Sonntag 1923: *P. troglodytes*); dorsal scapular nerve (Miller 1952: *P. paniscus*); C3 and C4 spinal nerves (Swindler & Wood 1973: *P. troglodytes*; these authors state that in modern humans it is also by C3 and C4).
- Notes: Sonntag (1923) described an origin in *P. troglodytes* from C1–C5 and Swindler & Wood (1973) from C1–C4, and an origin from C1–C4 was also found in one side of the bonobo described by Miller (1952), but in the other side of this specimen, as well as in one side of the *P. troglodytes* specimen dissected by Stewart (1936 and in the three *P. troglodytes* specimens reported by Schück (1913b) and the *P. troglodytes* specimens described by Hepburn (1892) the origin was from C1–C3. In the *P. troglodytes* specimen described by Gratiolet & Alix (1866) the origin was from C2–C3, while in one side of the *P. troglodytes*

specimen dissected by Stewart (1936) and in the *P. troglodytes* specimen analyzed by Champneys (1872) it was from C1–C2 only. In our specimens PFA 1016, PFA 1051 and HU PT1 the levator scapulae originated from C1–C3 (one slip per vertebra).
- Synonymy: Levator anguli scapulae (Wilder 1862, Macalister 1871, Champneys 1872, Hepburn 1892, Beddard 1893, Sonntag 1923, 1924); levator scapulae plus pars cranialis of serratus anticus (Schück 1913b); l'angulaire de l'omoplate (Gratiolet & Alix 1866, Broca 1869).

Levator claviculae (Figs. 8–10, 18)
- Usual attachments: From the atlas to the clavicle, passing deep to (covered either laterally or dorsally by) the trapezius (in *P. troglodytes* to, e.g., the lateral 1/3 of clavicle: Gratiolet & Alix 1866; the lateral 1/2 of clavicle: Champneys 1872; a point just lateral to the middle of the clavicle: Stewart 1936; in *P. paniscus* to the middle third of the clavicle: Miller 1952).
- Usual innervation: in one specimen of *P. troglodytes* it was C3 while in the other specimen of the same species it was C3 and C4 (Schück 1913b); branch of the anterior division of C3 through the cervical plexus (Miller 1952: *P. paniscus*).
- Notes: Miller (1952) and Andrews & Groves (1976) stated that in their chimpanzee specimens the levator claviculae passes lateral to the trapezius, but the reports of most authors, e.g., Champneys (1872), Schück (1913ab) and Ashton & Oxnard's (1963), agree with our observations (i.e., the levator claviculae is usually mainly deep to the trapezius). Broca (1869) stated that gorillas and chimpanzees lack a levator claviculae, but this is clearly not usually case in either of these taxa. Tyson (1699) stated that in the *P. troglodytes* specimen dissected by him the levator claviculae connected C2–C3 to the clavicle, but he was probably referring to part of the levator scapulae, because in the vast majority of chimpanzees and of other primates the muscle originates from the atlas only. Sutton (1883) stated that in the two *P. troglodytes* specimens dissected by him the levator claviculae is not present, but his statements should also be taken with caution, because he also stated that most of the facial muscles that are now recognized in chimpanzees (see above) were missing in his two specimens. The **atlantoscapularis posticus** (see Diogo et al. 2009a) is usually not present as a distinct muscle in chimpanzees.
- Synonymy: Tracheloclavicular (Wyman 1855); cléido-atloïdien or cléido-trachélien (Gratiolet & Alix 1866); omo-atlanticus (Macalister 1871); acromio-basilar or acromio-trachélien (Champneys 1872); omocervicalis, cleido-cervicalis, acromio-cervicalis, levator anticus scapulae (Barnard 1875, Schück 1913b); omo-trachelian (Sonntag 1923, 1924); atlantoscapularis anterior (Ashton & Oxnard 1963).

Subclavius (LSB 5.0 g; Fig. 18)
- Usual attachments: From the first rib to the clavicle.

- Usual innervation: Nerve to subclavius, from C5 and C6 (Hepburn 1892: *P. troglodytes*); nerve to subclavius, from brachial plexus (Miller 1952: *P. paniscus*); nerve to subclavius, from C5 (Swindler & Wood 1973: *P. troglodytes*).
- Notes: The **costocoracoideus** is usualy not present as a distinct muscle in chimpanzees, but these and most other primates do usually have a ligamentum costocoracoideum, which corresponds to the costocoracoideus muscle of primitive mammals such as monotremes (see Diogo et al. 2009a). However, as is the case in modern humans, a costocoracoideus muscle may be present as an anomaly in chimpanzees. For instance, Champneys (1872) describes a 'fibrous band from the coracoid to the sternum between the articulation of the clavicle and the first rib' in the *P. troglodytes* specimen dissected by him, which is partially fused to the subclavius muscle and the costocoracoid membrane, and which thus seems to correspond to the muscle costocoracoideus *sensu* the present study. Also, MacDowell (1910) describes a 'sterno-chondro-scapularis' in the *P. troglodytes* specimen dissected by him, which arises from a tendon inserted in rib 1 and inserts onto the coracoid process mesially to the attachment of the short head of the biceps brachii, the only muscular fibers of this 'sterno-chondro-scapularis' being a small group of fibers that occur at the coracoid attachment, and in a still smaller manner at the costal insertion; this 'sterno-chondro-scapularis' seemingly corresponds to a vestigial costocoracoideus *sensu* the present study.
- Synonymy: Sous-clavier (Gratiolet & Alix 1866).

Pectoralis major (LSB 587.0 g; Fig. 21)
- Usual attachments: The pars clavicularis runs from the medial portion of the clavicle and often from the sternum to the proximal humerus; the pars sternocostalis runs from the sternum and ribs to the proximal humerus, inserting proximal to the insertion of the pars clavicularis; the pars abdominalis runs from the ribs and aponeurosis of the external oblique, deep to the pars sternocostalis and pars clavicularis, to the proximal humerus and to the coracoid process of the scapula.
- Usual innervation: Anterior thoracic ('pectoral') nerve, which includes contributions from all of the trunks of the brachial plexus, or from the C7 and the combined trunk of C8 and T1 (Champneys 1872: *P. troglodytes*); medial and lateral pectoral nerves (Hepburn 1892, Swindler & Wood 1973: *P. troglodytes*; Miller 1952: *P. paniscus*).
- Function: Pars sternocostalis of pectoralis major exhibited moderate or high EMG activity during hoisting behavior and crutch-walking (Tuttle & Basmajian 1976, 1978b: *P. troglodytes*).
- Notes: The pectoralis major is sometimes, but not usually, blended with its counterpart in the midline (e.g., Loth 1931). According to the literature review of Andrews & Groves (1976) in *Pan* the clavicular origin of the clavicular head

is usually from the medial 1/4 to 1/2 of this bone, although it might be from the medial 2/3 in some cases (e.g., MacDowell 1910). In turn, according to the literature review of Gibbs (1999) there is no clavicular origin of the pectoralis major in 2/13 of *Pan*. Regarding the origin from the ribs, its posterior limit is often on the fourth, fifth or sixth ribs according to the literature review of Andrews & Groves (1976), although in some cases it might extend to rib 8 (e.g., Gibbs 1999). The **'tensor semi-vaginae articulationis humero-scapularis'** *sensu* Macalister (1871) and Huntington (1903), which is also designated as **'sterno-humeralis'**, **'sterno-chondro-humeralis'** or **'pectoralis minimus'** by these authors, is a muscle that is occasionally present in modern humans, and that is found on the left side in a *P. troglodytes* specimen described by Macalister (1871), arising from the cartilages of ribs 1–2 (for more details on this structure and its evolution and homologies, see Diogo & Wood 2012). MacDowell (1910) stated that in the *P. troglodytes* specimen dissected by him there is an 'additional muscle chondro-epitrochlearis' that lies deep (dorsal) to the sternocostal head of the pectoralis major, originates from ribs 2–4 and the fascia underlying this head of the pectoralis major, and runs parallel with this head to insert with it onto the humerus. The structure MacDowell designates as **'chondro-epitrochlearis'** is most likely an additional slip of the pectoralis major, and it does not correspond to the anomaly/variant seen in modern humans referred to as **'muscle chondroepitrochlearis'**. In fact, in modern humans this variant/anomalous muscle usually runs from the ribs to the median intermuscular septum or onto the medial epicondyle of the humerus (hence its name 'chondroepitrochlearis'). The **sternalis**, **'pectoralis quartus'** and **panniculus carnosus** are usually not present as distinct muscles in chimpanzees.
- Synonymy: Pectoralis major plus the lower portion of pectoralis minor (Hartmann 1886); pectoralis major plus pectoralis abdominis, abdominalis, and/or chondroepitrochlearis quartus (MacDowell 1910, Gibbs 1999).

Pectoralis minor (LSB 52.0 g; Figs. 18–19, 25)
- Usual attachments: From the proximal humerus and/or the glenohumeral joint capsule and, sometimes, from the coracoid process of the scapula, to the ribs (e.g., ribs 2–4: Hepburn 1892, Sonntag 1923, Stewart 1936, Miller 1952, and our specimens PFA 1016 and PFA UNC; ribs 2–3 and aponeurosis from ribs 4: Champneys 1872, Ziegler 1964; ribs 2–5: Gratiolet & Alix 1866; ribs 2–3: Chapman 1879, and our specimen PFA 1077; ribs 3–5: Sutton 1883, Swindler & Wood 1973, and our specimen PFA 1051; ribs 1–4: MacDowell 1910).
- Usual innervation: Anterior thoracic (pectoral) nerve, which includes elements from all three trunks of the brachial plexus, or from C7 and the combined trunk of C8 and T1 (Champneys 1872: *P. troglodytes*); medial pectoral nerve (Hepburn 1892, Swindler & Wood 1973: *P. troglodytes*; Miller 1952: *P. paniscus*).

- Notes: In some of the *Pan* specimens described by Gratiolet & Alix (1866), Humphry (1867), Macalister (1871), Barnard (1875), Hartmann (1886), Beddard (1893), Lander (1918), Miller (1952) and Andrews & Groves (1976) and dissected by us, some fibers of the pectoralis minor may insert onto the coracoid process, but even in these cases the great majority of the fibers of this muscle insert onto other structures (e.g., the capsule of the glenohumeral joint). Moreover, in most of the *Pan* specimens described (e.g., Wyman 1855, Huxley 1864, Broca 1869, Champneys 1872, Chapman 1879, Sutton 1883, Hepburn 1892, MacDowell 1910, Sonntag 1923, 1924, Miller 1952, Ziegler 1964, Swindler & Wood 1973) there is no attachment at all onto the coracoid process. In a few *Pan* specimens, such as one specimen described by Wilder (1892), the pectoralis minor inserted onto the coracoid process on one side of the body, while on the other side of the body it inserted onto the greater tuberosity of the humerus instead. In summary, the usual condition for *Pan* is for there to be no direct attachment onto the coracoid process (e.g., Ashton & Oxnard 1963 suggest three-quarters of *Pan* specimens there is effectively no insertion onto the coracoid process). The **pectoralis tertius ('xiphihumeralis')** (see Diogo et al. 2009a) is usually not present as a distinct muscle in chimpanzees.
- Synonymy: Upper portion of pectoralis minor (Hartmann 1886).

Infraspinatus (LSB 298.0 g; Figs. 24, 26)
- Usual attachments: From the infraspinous fossa of the scapula and the infraspinatus fascia to the greater tuberosity of the humerus and in some cases to the capsule of the glenohumeral joint (e.g., Beddard 1893, Miller 1952).
- Usual innervation: Suprascapular nerve, from C5 and C6 (Hepburn 1892: *P. troglodytes*); suprascapular nerve (Miller 1952: *P. paniscus*; Swindler & Wood 1973: *P. troglodytes*).
- Function: Infraspinatus exhibited EMG activity during arm-raising; low or nil activity was exhibited during hoisting behavior (Tuttle & Basmajian 1978a: *P. troglodytes*). EMG study indicated that the supraspinatus, infraspinatus and subscapularis have "completely different" functions; the supraspinatus acts as a more or less pure abductor of the arm, assisting the deltoideus, while the infraspinatus is more an abductor/lateral rotator and the different parts of the subscapularis have distinct functions such as abduction/medial rotation and adduction/medial rotation (Larson & Stern 1986: *P. troglodytes*).
- Synonymy: Sous-épineux (Gratiolet & Alix 1866).

Supraspinatus (LSB 125.0 g; Figs. 24, 26)
- Usual attachments: From the supraspinous fossa of the scapula and the supraspinatus fascia to the greater tuberosity of the humerus and in many cases to the capsule of the glenohumeral joint, as well as, in a few cases, to the acromioclavicular ligaments (e.g., Gratiolet & Alix 1866).

- Usual innervation: Suprascapular nerve, from C5 and C6 (Hepburn 1892: *P. troglodytes*); suprascapular nerve (Miller 1952: *P. paniscus*; Swindler & Wood 1973: *P. troglodytes*).
- Function: Supraspinatus exhibited high or moderate EMG activity during arm-raising, quiet tripedal and quadrupedal stance, and, unlike *Gorilla* and *Pongo*, also during hoisting behavior (Tuttle & Basmajian 1978a: *P. troglodytes*). For details about EMG studies of *P. troglodytes* done by Larson & Stern (1986) see infraspinatus.
- Notes: The muscle **scapuloclavicularis,** occasionally present in modern humans, has not been described in chimpanzees nor was it found in the chimpanzees dissected by us. Potau et al. (2009) studied the relative masses of the deltoideus, subscapularis, supraspinatus, infraspinatus and teres minor in chimpanzees, orangutans and modern humans and showed that for all of the proportional values in *Pongo pygmaeus* there is marked overlap between the ranges of values in the two species. According to Potau et al. this may indicate that the functional requirements of the glenohumeral joint are similar in a great ape with fundamentally suspension/vertical climbing locomotion and in a bipedal great ape that uses the upper extremity for essentially manipulative functions. They also showed that chimpanzees have a proportionally larger mass of the rotator cuff versus the deltoideus, primarily because of the greater role played by the subscapularis. The latter may be an adaption to knuckle-walking because in this type of locomotion the subscapularis muscle acts as the principal stabilizer of the glenohumeral joint, compensating for the shearing force on this joint caused by the dorsal position of the scapula. Modern humans have a larger mass of the deltoideus versus that of the subscapularis; the deltoideus is the principal elevator muscle of the upper extremity in the scapular plane, but its contraction elevates the humeral head to the undersurface of the acromion. This may compress and injure the supraspinatus tendon, but is usually protected by the simultaneous contraction of the rotator cuff muscles. The smaller proportional mass of the rotator cuff musculature, particularly the subscapularis, in modern humans versus chimpanzees, together with the larger proportional mass of the deltoid in modern humans, may explain the 1-3 mm upward movement of the humeral head during the first 30° to 60° of elevation of the upper extremity in the scapula plane in modern humans; this slight movement pinches the supraspinatus tendon between the humeral head and acromion. This anatomical pattern in modern humans may provide the structural explanation for the tendency of certain individuals to suffer a range of subacromial syndromes (e.g., sports that require elevation of the upper extremity).
- Synonymy: Sus-épineux (Gratiolet & Alix 1866).

Deltoideus (LSB 576.0 g; Figs. 18–19)
- Usual attachments: From the lateral portion of the clavicle (pars clavicularis —either the lateral 1/2: Gratiolet & Alix 1866, Beddard 1893, Sonntag 1923; the lateral 1/3: Miller 1952, Ashton & Oxnard 1963, Swindler & Wood 1973; or more often the lateral 1/4: Andrews & Groves 1976), the acromion (pars acromialis) and the spine of the scapula and the infraspinatus fascia (pars spinalis) to the humerus.
- Usual innervation: Axillary (circumflex) nerve (Hepburn 1892, Sonntag 1923, Swindler & Wood 1973: *P. troglodytes*; Miller 1952: *P. paniscus*).
- Function: Deltoideus exhibited moderate to high EMG activity during arm-raising—no data for hoisting behavior (Tuttle & Basmajian 1978a: *P. troglodytes*); EMG indicated that the 'anterior and middle' portions of the deltoid are mainly related with the elevation of the arm, while the 'posterior' portion of this muscle has a "completely different" function (e.g., humeral retraction Larson & Stern 1986: *P. troglodytes*).

Teres minor (LSB 52.0 g; Figs. 24, 26)
- Usual attachments: From the infraspinatus fascia and the lateral border of the scapula to the greater tuberosity of the humerus, and often also to the humeral shaft just distal to the same tuberosity (e.g., Gratiolet & Alix 1866, Sonntag 1923; Swindler & Wodd 1973; our specimens, e.g., PFA 1016, PFA 1051, HU PT1).
- Usual innervation: Axillary (circumflex) nerve (Champneys 1872, Hepburn 1892, Sonntag 1923, Swindler & Wood 1973: *P. troglodytes*; Miller 1952: *P. paniscus*).
- Function: EMG study indicated that the teres minor acted more as the 'posterior' portion of the deltoideus and, even, as the teres major, being, e.g., related with functions such as humeral retraction/adduction, as is often the case in orangutans and in modern humans (contrary to what is often stated in textbooks of human anatomy; Larson & Stern 1986: *P. troglodytes*).
- Synonymy: Petit rond (Gratiolet & Alix 1866).

Subscapularis (LSB 395.0 g; Figs. 19, 24)
- Usual attachments: From the subscapular fossa of the scapula to the lesser tuberosity of the humerus, and often also to the humeral shaft just distal to this tuberosity (e.g., Sonntag 1923; our specimens, e.g., PFA 1016, PFA 1051, PFA 1077, PFA UNC).
- Usual innervation: Subscapular nerves (Hepburn 1892: *P. troglodytes*; Miller 1952: *P. paniscus*; upper and lower subscapular nerves (Swindler & Wood 1973: *P. troglodytes*).
- Function: Moderate or high EMG activity during arm-raising; low or nil activity was exhibited during hoisting behavior (Tuttle & Basmajian 1978a: *P. troglodytes*); for details of EMG study of *P. troglodytes* by Larson & Stern (1986), see infraspinatus.

- Notes: In the *P. troglodytes* specimen dissected by Ziegler (1964) the subscapularis has an accessory head arising from the superior half-inch of the axillary border of the bulk of the muscle—a few fibers of this head insert onto the capsular ligament while the rest attach along a half-inch strip of the humerus distal to the lesser tuberosity.

Teres major (LSB 264.0 g; Figs. 24, 26)
- Usual attachments: From the lateral border and inferior angle of the scapula to the proximal portion of the intertubercular groove of the humerus by means of a distal tendon that is usually not fused to the distal tendon of the latissimus dorsi.
- Usual innervation: Subscapular nerves (Champneys 1872, Hepburn 1892: *P. troglodytes*; Miller 1952: *P. paniscus*); lower subscapular nerve (Swindler & Wood 1973: *P. troglodytes*).
- Function: EMG activity during hoisting behavior (Tuttle & Basmajian 1976: *P. troglodytes*).
- Notes: According to Gibbs' (1999) review of the literature, a complete separation between the insertion tendons of the latissimus dorsi and of the teres major is found in 5/8 *Pan* (e.g., found by Ziegler 1964, MacDowell 1910, Stewart 1936, and other authors, but not reported by Gratiolet & Alix 1866, Sonntag 1923, Miller 1952, Hepburn 1892, Champneys 1872, and Dwight 1895, who referred to a partial fusion of the tendons). If we add Gibbs' (1999) numbers with the numbers found in our own dissections (no fusion of the distal tendons of the two muscles in 4 of the 7 chimpanzees where we analyzed this feature), the total numbers are: no fusion between the latissimus dorsi and the teres major in 9/15 *Pan*. Thus, contrary to the condition in hylobatids and orangutans (see below) there is often no fusion in *Pan* and there is usually no fusion in gorillas and modern humans (Diogo & Wood 2011, 2012).
- Synonymy: Grand rond (Gratiolet & Alix 1866).

Latissimus dorsi (LSB 489.0 g; Figs. 19, 26)
- Usual attachments: From the vertebrae, ribs, thoracolumbar fascia and, often, directly and/or indirectly from the pelvis, to the intertubercular groove of the humerus; in a *P. paniscus* specimen reported by Miller (1952) the muscle also originates from the inferior angle of the scapula.
- Usual innervation: thoracodorsal nerve (Champneys 1872, Hepburn 1892: *P. troglodytes*; Miller 1952: *P. paniscus*); C7, C8 (Swindler & Wood 1973: *P. troglodytes*).
- Function: EMG activity during hoisting behavior (Tuttle & Basmajian 1976: *P. troglodytes*).
- Synonymy: Grand dorsal (Gratiolet & Alix 1866, Broca 1869).

Dorsoepitrochlearis (LSB 51.0 g; Figs. 19, 24, 26–28)
- Usual attachments: From the distal portion of the latissimus dorsi and occasionally from the coracoid process of the scapula (e.g., Gratiolet & Alix 1866) to the medial epicondyle of the humerus and occasionally also to the adjacent supracondylar ridge of the humerus; in a few cases the dorsoepitrochlearis attachment extends distally to attach directly onto the olecranon process of the ulna (see Notes below).
- Usual innervation: Radial nerve (Champneys 1872, Hepburn 1892, Swindler & Wood 1973: *P. troglodytes*; Miller 1952: *P. paniscus*); branch of radial nerve that anastomoses with a similar branch of this nerve passing into the medial side of the medial head of the triceps brachii (Ziegler 1964: *P. troglodytes*).
- Function: A minor part of the action of the muscle may be extension of the forearm (Ziegler 1964: *P. troglodytes*).
- Notes: As noted by Aiello & Dean (1990), in non-human hominoids the dorsoepitrochlearis is usually mainly attached onto the medial epicondyle, the intermuscular septum and/or surrounding structures, but not onto the olecranon process or the olecranon fascia. Regarding *Pan*, Tyson (1699), Gratiolet & Alix (1866), Barnard (1875), Hartmann (1886), Beddard (1893), Dwight (1895), MacDowell (1910), Schück (1913a), Sonntag (1923), Miller (1932, 1952), Swindler & Wood (1973), and Payne (2001) found a bony insertion onto the medial epicondyle only, as we did in our specimens, while Ashton & Oxnard (1963) refer to insertions onto the medial epicondyle and the supracondylar ridge of the humerus, and only very few authors, e.g., Vrolik (1841), Testut (1883), Grönroos (1903) and Ziegler (1964), described a bony insertion onto both the medial epicondyle and olecranon process of the ulna. According to the literature review done by Gibbs (1999), the usual condition for *Pan* is effectively that in which there is a bony insertion onto the medial epicondyle only. MacDowell (1910) described a *P. troglodytes* specimen in which the dorsoepitrochlearis received a fasciculus from the coracobrachialis.
- Synonymy: Appendix of the latissimus dorsi (Tyson 1699); muscle accessoire du long dorsal (Gratiolet & Alix 1866, Broca 1869); latissimo-condylus or latissimo-epitrochlearis (Barnard 1875); latissimo-condyloideus (Chapman 1879; Hepburn 1892, MacDowell 1910, Pira 1913); latissimo-tricipitalis (Schück 1913a,b, Fick 1925).

Triceps brachii (LSB 652.0 g; Figs. 24, 26–27, 30)
- Usual attachments: From at least half of the length of the lateral border of the scapula (caput longum) and from the shaft of the humerus (caput laterale and caput mediale) to the olecranon process of the ulna.
- Usual innervation: Radial nerve (Hepburn 1892, Sonntag 1923, Swindler & Wood 1973: *P. troglodytes*; Miller 1952: *P. paniscus*); radial nerve but also by a branch of the ulnar nerve in at least some specimens (Champneys 1872: *P. troglodytes*).

- Notes: In *Pan* the long head of the triceps brachii usually originates from at least half of the length of the lateral border of the scapula. This was noted by Loth (1931). Loth (1931), Ziegler (1964) and Swindler & Wood (1973) and we also found this to be the case in the specimens examined by us except in the PFA 1077 and PFA UNC infants, in which the attachment was to one third of the length of the lateral border of the scapula, and Gratiolet & Alix (1886) referred to the whole lateral border; only Sonntag (1923 referred to one-quarter). It should be noted that the triceps brachii of modern humans includes the **articularis cubiti,** which is listed in Terminologia Anatomica 1998 as a muscle that is usually present in modern humans, but this term refers to a subdivision of the triceps brachii that runs from the main body of that muscle to the posterior aspect of the capsule of the elbow joint, thus lifting the capsule away from the joint; it should not to be confused with the muscle anconeus (see below). An articularis cubiti bundle of the triceps brachii has also been described in a few chimpanzees (e.g., Champneys 1872 and Sonntag 1923), but it was given the name 'subanconeus'.
- Synonymy: Triceps plus subanconeus (Champneys 1872, Sonntag 1923); vaste externe plus vaste interne plus scapulo-olécrânien, which correspond respectively to the caput laterale, caput mediale and caput longum *sensu* the present study (Gratiolet & Alix 1866); multiceps extensor cubiti (Barnard 1875); triceps extensor cubiti (Hepburn 1892).

Brachialis (LSB 243.0 g; Figs. 27–28)
- Usual attachments: From the humeral shaft (i.e., distal to the surgical neck of the humerus) to the ulnar tuberosity.
- Usual innervation: Median nerve (Gratiolet & Alix 1866, Sutton 1883, Bolk 1902: *P. troglodytes*); musculocutaneous nerve (Hepburn 1892, Sonntag 1923, Swindler & Wood 1973, Koizumi & Sakai 1995: *P. troglodytes*; Miller 1952: *P. paniscus*).
- Notes: Ziegler (1964) stated that on the right side of the *P. troglodytes* specimen dissected by him two fleshy slips left the center of the anterior aspect of the main body of the brachialis and ran distally to join the posterior aspect of the biceps brachii near its tendon of insertion; according to him such slips are recorded as rare anomalies in modern humans. Howell & Straus (1932) stated that on both sides of the *P. troglodytes* specimen dissected by them the medial belly of the brachialis was partially separated from the lateral belly, the medial belly having the more extensive origin from the distal half of the humerus and its medial epicondyle, inserting in the form of a tendon onto the ulna just distal to the coronoid process; the lateral belly had a shorter origin onto the humerus near the deltoid insertion. Our infant specimen PFA 1077 has a configuration similar to that shown in plate 1 of Howell & Straus 1932 (i.e., proximally the brachialis is partially divided into medial and lateral bundles, but the bundles fuse distally and insert together onto the ulnar tuberosity.

- Synonymy: Brachial antérieur (Gratiolet & Alix 1866); brachialis anticus (Champneys 1872, Hepburn 1892, Beddard 1893, Sonntag 1923).

Biceps brachii (LSB 364.0g; Figs. 19, 24–25, 27–28, 30)
- Usual attachments: From the proximal humerus (caput breve) and the supraglenoid tubercle of the scapula (caput longum), to the bicipital tubercle of the radius (common tendon) and to fascia covering the forearm flexors (bicipital aponeurosis, which in chimpanzees is usually not fleshy, i.e., it does not form a 'lacertus carnosus': see Notes below).
- Usual innervation: Median nerve (Gratiolet & Alix 1866, Sutton 1883, Bolk 1902: *P. troglodytes*); musculocutaneous nerve (Hepburn 1892, Sonntag 1923, Swindler & Wood 1973, Koizumi & Sakai 1995: *P. troglodytes*; Miller 1952: *P. paniscus*).
- Notes: In hominoids such as *Hylobates*, *Gorilla*, *Pan* and modern humans, the biceps brachii is usually prolonged distally by a bicipital aponeurosis ('lacertus fibrosus' in *Gorilla*, *Pan* and modern humans; often 'lacertus carnosus' in *Hylobates*) which is commonly associated with the fascia covering forearm muscles such as the pronator teres. Regarding *Pan*, almost all, if not all, specimens have a bicipital aponeurosis, as reported by Gratiolet & Alix (1866), Sonntag (1923), Loth (1931), Howell & Straus (1932), Glidden & De Garis (1936), Miller (1952), Ziegler (1964) and Swindler & Wood (1973), and corroborated by our dissections. Plate 1 of Howell & Straus (1932) shows a chimpanzee specimen with 'accessory heads of the biceps brachii' (in addition to the short and long heads and to the bicipital aponeurosis). On the right side of the *Pan* sp. specimen dissected by Howell & Straus they report an 'accessory head' of the long head originating from the capsule of the gleno-humeral joint and fusing distally with the main body of the long head below the head of the humerus. On the left side of the same specimen Howell & Straus report the presence of two 'accessory heads' of the long head.
- Synonymy: Biceps flexor cubiti (Owen 1868, Hepburn 1892); scapulo-radial plus coraco-antebrachial, which correspond respectively to the caput longum and caput breve *sensu* the present study (Gratiolet & Alix 1866).

Coracobrachialis (LSB 87.0 g; Figs. 19, 24, 27)
- Usual attachments: From the coracoid process of the scapula to the proximal portion of the humerus, being usually (e.g., Macalister 1871, Champneys 1872, Dwight 1895, Sonntag 1923; our dissections), but not always (e.g., Tyson 1699), pierced by the musculocutaneous nerve
- Usual innervation: Median nerve (Gratiolet & Alix 1866, Sutton 1883, Bolk 1902: *P. troglodytes*); musculocutaneous nerve (Hepburn 1892, Swindler & Wood 1973, *P. troglodytes*; Miller 1952: *P. paniscus*); musculocutaneous nerve, but occasionally also median nerve (Sonntag 1923: *P. troglodytes*); musculocutaneous nerve, except on one side of a specimen (Koizumi & Sakai 1995: *P. troglodytes*).

- Notes: The observations of most authors, as well as our dissections, show that the usual condition for chimpanzees is that the **coracobrachialis superficialis/longus** and **coracobrachialis profundus/coracobrachialis brevis** are not present as distinct structures, i.e., the coracobrachialis has a single bundle that corresponds to the **coracobrachialis medius/coracobrachialis proprius** of other mammals. A few exceptions follow. Macalister (1871) reported a *P. troglodytes* specimens that had a "small rudiment" of the coracobrachialis profundus/brevis; Parsons (1898ab) stated that small rudiment is found in 30% of chimpanzees and Howell & Straus (1932) stated that it is present in 4 out of 13 chimpanzees. Howell & Straus (1932) noted, correctly, that a coracobrachialis superficialis/longus is only found in a few mammals, and stated that they found it as an anomaly on the right side of the *Pan* specimen dissected by them. According to their report, the coracobrachialis superficialis/longus originates together with the short head of the biceps brachii and continues superficial to all of the nerves and muscles of the arm, except for the dorsoepitrochlearis, to its insertion on the medial epicondyle (see their plate 1). The musculocutaneous nerve passes between it and the coracobrachialis medius/proprius, which originates from the coracoid process and inserts on the humerus. According to these authors no part of the coracobrachialis can be considered as representing a coracobrachialis profundus/brevis; on the left side of their specimen; there is only a coracobrachialis medius/proprius that originated from the coracoid process and then passed superficial (its medial portion) and deep (its lateral portion) to the musculocutaneous nerve. A distinct coracobrachialis longus/superficialis running from the coracoid process of the scapula to the medial epicondyle of the humerus was also reported in one of the four *P. troglodytes* dissected by Oishi et al. 2009. The 'coracobrachialis longus'—or 'superficialis'—reported by Sonntag 1923 does not correspond to the coracobrachialis superficialis/longus *sensu* the present study, i.e., this reports is most likely due to the use of an erroneous nomenclature, because the coracobrahialis superficialis/longus is a bundle that should be superficial to almost all the nerves and muscles of the arm (e.g., Diogo & Abdala 2010 and Diogo & Wood 2012).

Pronator quadratus (LSB 17.4 g; Fig. 28)
- Usual attachments: From the distal portion of the ulna to the distal portion of radius.
- Usual innervation: Anterior (volar) interosseous branches of the medial nerve (Miller 1952: *P. paniscus*).
- Function: According to Tuttle (1969) there is a trend towards reducing the pronator musculature in the African apes, which may be associated with the fact that they do not engage as frequently as orangutans in arboreal movements that require a wide range of supination and pronation of the forearm; furthermore,

the African apes, particularly *Gorilla*, require considerable stability at the elbow joint to maintain the fully extended forearm against the compressive forces incurred during knuckle-walking.
- Notes: In the *P. troglodytes* specimen dissected by Dwight (1895) the pronator quadratus has two layers that are only partly separable, the more superficial beginning as a tendon at the top of the ulnar origin and expanding as it passes downward and outward across the muscle; it gradually becomes muscular, and having reached the outside of the radius, runs distally to the ridge of the trapezium. The deeper, more transverse part of the pronator quadratus runs obliquely and distally towards the radius; a few radial fibers go to the anterior aspect of the wrist capsule.
- Synonymy: Carré pronateur (Gratiolet & Alix 1866).

Flexor digitorum profundus (LSB 315.3 g; Figs. 28–29, 31–33, 37–38, 40)
- Usual attachments: From the radius, ulna, and interosseous membrane to the distal phalanges of digits 2, 3, 4 and 5 and, sometimes, to digit 1 through a tendon that is often thin and/or not continuous to the main body of the muscle (see Notes below).
- Usual innervation: Median (usually to digits 2 and 3, and digit 1 if there is a portion of the muscle going to this digit) and ulnar (usually to digits 4 and 5) nerves (Champneys 1872, Hepburn 1892, Swindler & Wood 1973: *P. troglodytes*; Miller 1952: *P. paniscus*).
- Notes: Among the primates dissected by us, hylobatids and modern humans are the only ones in which the flexor pollicis longus is usually present as a distinct, independent muscle (e.g., the **flexor pollicis longus** is usually not present as a distinct muscle in chimpanzees Diogo & Wood 2011, 2012). That is, when some authors state that there is a 'flexor pollicis longus' in a chimpanzee they are either referring to the part of the flexor digitorum profundus that goes to digit 1, or to the belly of this muscle that often goes to both digits 1 and 2, and not really to a distinct muscular belly going exclusively to digit 1 as that usually found in hylobatids and modern humans (see Synonymy below). In fact, in *Pongo*, *Gorilla* and *Pan* the tendon to digit 1 is effectively often absent or vestigial, as corroborated in the specimens dissected by us (e.g., Figs. 32, 33, 37, 38 and 40) and by others, except in a few specimens of these genera according to Straus (1942a) and in the *P. paniscus* specimen described by Miller (1952), in which the tendon to digit 1 is said to be similar to the other tendons of the flexor digitorum profundus. According to the literature reviews of Keith (1894b) and Gibbs (1999) there is no tendon of the flexor digitorum profundus to digit 1 in 10/25 chimpanzees and in 13/43 chimpanzees, respectively. Straus (1942b) compiled evidence from his own dissections of hominoids and from data available on the literature and stated that within 47 chimpanzees the tendon of the flexor digitorum profundus to digit 1 is completely absent in

30% (14 of 47), is functionless in 22% (10.5 of 47), and is in direct functional continuity with the radial muscle belly of the flexor digitorum profundus in 48% (22 of 47); thus the flexor digitorum profundus tendon to digit 1 is entirely without function in at least half of the members of *Pan* (52% of chimpanzees).
- Synonymy: Flexor communis profundus (Wyman 1855); fléchisseurs des troisièmes phalanges or fléchisseurs profonds or fléchisseurs perforants (Gratiolet & Alix 1866); fléchisseur profond des doigts (Broca 1869); flexor profundus et pollicis (Macalister 1871); flexor profundus digitorum plus flexor longus pollicis (Champneys 1872, Beddard 1893, Sonntag 1923); flexor profundus digitorum plus flexor profundus indicis plus flexor longus pollicis (Sutton 1883); flexor digitorum communis profundus (Barnard 1875); flexor digitorum profundus plus flexor pollicis longus (Swindler & Wood 1973).

Flexor digitorum superficialis (LSB 275.3 g; Figs. 27–29, 31, 39–40)
- Usual attachments: From the radius (caput radiale), the ulna and the medial epicondyle of the humerus (caput humeroulnare) to the middle phalanges of digits 2–5.
- Usual innervation: Median nerve (Hepburn 1892, Sonntag 1923, Swindler & Wood 1973: *P. troglodytes*; Miller 1952: *P. paniscus*).
- Function: According to Tuttle (1969) in great apes the fasciculi of the flexor digitorum superficialis and/or flexor digitorum profundus to the individual digits, especially digit 2, frequently appear as entities in the middle of the forearm and may be traced proximally at least to this level by following the individual tendons of the flexor digitorum superficialis/profundus muscles. According to Tuttle this is probably related to the ability to flex the fingers independently. This is of considerable advantage to orangutans and chimpanzees in that it allows the animals to hold a number of small twigs with some digits while the remaining fingers reach for additional supports. If some of the twigs break, the individual does not have to open the whole hand in order to grasp replacements. Also according to Tuttle, it is probable that the flexor digitorum superficialis/profundus muscles are active to prevent extreme hyperflexion at the metacarpophalangeal joints in *Gorilla* and *Pan* during static knuckle-walking postures; this contrasts with the conclusions of Ziegler (1964) that the major danger of 'collapse' at these joints was in the direction of flexion).
- Notes: Sutton (1883), Hepburn (1892), Dwight (1895), MacDowell (1910), Sonntag (1923), Loth (1931), Miller (1952), Jouffroy (1971), Swindler (1973) and Gibbs (1999) refer to an origin of the flexor digitorum superficialis in chimpanzees that includes the ulna, radius and the medial epicondyle of the humerus and/or the common flexor tendon. Such an origin was also found in our dissections. Only a few authors report examples in which there is no ulnar origin (e.g., Beddard 1893) and/or no radial origin (e.g., Gratiolet & Alix 1866).

- Synonymy: Fléchisseurs des secondes phalanges or fléchisseurs superficiels or fléchisseurs perforés (Gratiolet & Alix 1866); flexor sublimis digitorum (Macalister 1871, Champneys 1872, Hepburn 1892, Beddard 1893, Dwight 1895, Sonntag 1923); flexor sublimis digitorum plus flexor sublimis indicis (Sutton 1883); flexor digitorum sublimis (MacDowell 1910, Miller 1952).

Palmaris longus (LSB 15.5 g; Figs. 27, 28)
- Usual attachments: Medial epicondyle of the humerus to the palmar aponeurosis.
- Usual innervation: Median nerve (Hepburn 1892, Ziegler 1964, Swindler & Wood 1973: *P. troglodytes*; Miller 1952: *P. paniscus*).
- Notes: According to Ziegler 1964, the branch of the median nerve that innervates the palmaris longus in the *P. troglodytes* specimen dissected by him pierces the flexor digitorum superficialis before entering the palmaris longus; this might suggest that the palmaris longus of at least some primates such as chimpanzees derives phylogenetically from the flexor digitorum superficialis (for more details, see Diogo & Abdala 2010 and Diogo & Wood 2012). In the *Pan* specimens described by Tyson (1699; 1 specimen), Vrolik (1841; 1 specimen), Wilder (1862; 1 specimen), Gratiolet & Alix (1866; 1 specimen), Humphry (1867; 2 specimens), Macalister (1871; 1 specimen), Champneys (1872; 1 specimen), Chapman (1879; 1 specimen), Hepburn (1892; 1 specimen), Beddard (1893; 1 specimen), Dwight (1895; 1 specimen), MacDowell (1910; 1 specimen), Ribbing & Hermansson (1912; 1 specimen), Miller (1952; 1 specimen), Ziegler (1964; 1 specimen), Swindler & Wood (1973; 1 specimen), Sarmiento (1994; 2 specimens), Ogihara et al. (2005; 1 specimen), Oishi et al. (2009; 4 specimens) and Kikuchi (2010a; 1 specimen) the muscle was present, but it was absent in the specimen dissected by Sonntag (1923), in 1 of the 3 specimens reported by Sonntag (1924), in one of the four upper limbs dissected by Carlson (2006), and in 1 out of the 6 specimens dissected by us in which we could discern this feature in detail. The palmaris longus is only present in 9 out of 12 chimpanzee upper limbs according to Keith's (1899) review, in 95% of the cases according to Loth's (1931) review, in 15.5 out of 17 cases according to Sarmiento's (1994) review, and in 19 out of 28 cases according to Gibbs' (1999) review.
- Synonymy: Fléchisseur des premières phalanges or palmaire grêle (Gratiolet & Alix 1866).

Flexor carpi ulnaris (LSB 109.3 g; Figs. 27, 28)
- Usual attachments: From the ulna (caput ulnare) and usually also from the medial epicondyle of the humerus (caput humerale) to the pisiform, and sometimes also to the base of metacarpal V (e.g., Gratiolet & Alix 1866, Miller 1952) and/or to the hamate (e.g., our specimen PFA 1077).
- Usual innervation: Ulnar nerve (Champneys 1872, Hepburn 1892, Sonntag 1923, Swindler & Wood 1973: *P. troglodytes*; Miller 1952: *P. paniscus*).

- Notes: A bony origin from the ulna and humerus is usually present in chimpanzees and was found in the *Pan* specimens dissected by us and described by Gratiolet & Alix (1866), Hepburn (1892), Beddard (1893), Sonntag (1923), Miller (1952) and Swindler & Wood (1973).
- Synonymy: Métacarpien palmaire du cinquième doigt or cubital antérieur or cubital palmaire (Gratiolet & Alix 1866, Jouffroy & Lessertisseur 1957).

Epitrochleoanconeus (LSB 2.1 g; Fig. 27)
- Usual attachments: Medial epicondyle of the humerus to the olecranon process of the ulna.
- Usual innervation: Ulnar nerve (Howell & Straus 1932, Ziegler 1964: *P. troglodytes*; Miller 1952: *P. paniscus*).
- Notes: In *Pan* the epitrochleanconeus is usually present as a distinct muscle. There are some cases in which the muscle is not described, but it is not clear if this is because it is absent. Some authors (e.g., Macalister 1871—1 specimen; Sonntag 1923—1 specimen) comment when a muscle is absent, but it is more common for authors to state that the muscle was present, e.g., in the specimen described by Gratiolet & Alix 1866, on one side of the body of the specimen dissected by Miller 1952 (absent on the other side), on one side of the body of the specimen described by Ziegler 1964 (he could not discern if the muscle was present or not on the other side), in the specimen illustrated by Swindler & Wood 1973 (in their p. 327 they state that this muscle is absent in *Pan*, but on p. 146 they state that it is present and they clearly show it in the specimen illustrated in a figure in their p. 147), and in 2 of the 5 specimens dissected by us in which we could discern if the muscle was present or not (in the other 3 infant specimens the muscle seemed to be undifferentiated). It should be noted that the epitrochleoanconeus should be, by definition, as shown in plate 71 of Swindler & Wood (1973) (i.e., it should be a small muscle that derives from the anlage that gives rise to the flexor carpi ulnaris and it should run mainly from the medial epicondyle of the humerus to the olecranon process of the ulna. It is not clear if the muscle shown in the chimpanzee of plate 1 of Howell & Straus (1932) is an epitrochleoanconeus, because this muscle can be seen in a ventral view of the upper limb (and usually it should only be seen from a dorsal view), because it is rather well-developed and not small, and also because the muscle does not seem to insert onto the olecranon process of the ulna.
- Synonymy: Anconeus quartus, anconeus medialis, anconeus sextus, anconeus parvus, or tensor fasciae antebrachii (Jouffroy 1971).

Flexor carpi radialis (LSB 107.3 g; Fig. 27)
- Usual attachments: From the medial epicondyle of the humerus and usually also from the radius, to the base of metacarpal II and usually also to the base of metacarpal III.
- Usual innervation: Median nerve (Miller 1952: *P. paniscus*).

- Notes: In *Pan* the muscle has bony origins from at least the humerus and the radius. This condition was found in the *Pan* specimens described by Gratiolet & Alix (1866), Hartmann (1886), Dwight (1895), Sonntag (1923), Miller (1952), Ziegler (1964) and Swindler & Wood (1973) and dissected by us; there are very few descriptions of chimpanzees that refer to a bony origin from the humerus only (e.g., Hepburn 1892 and Beddard 1893). Also, in *Pan* the muscle often inserts onto both metacarpals II and III. In the survey of Gibbs (1999) this author stated that the muscle is inserted onto both metacarpals II and III in 5 out of 10 chimpanzees, but Jouffroy (1971) argue that an insertion onto both metacarpals is the usual condition for *Pan*. An exclusive insertion onto metacarpal II was described by Gratiolet & Alix (1866), Beddard (1893), Dwight (1895) and Miller (1952), but in the specimens dissected by Hepburn (1892), Sonntag (1923), Ziegler (1964) and Swindler & Wood (1973), as well as in the three specimens dissected by us in which we could discern this feature in detail, the insertion was onto both metacarpals II and III.
- Synonymy: Métacarpien palmaire du deuxiéme doigt or radial antérieur ou grand palmaire (Gratiolet & Alix 1866, Jouffroy & Lessertisseur 1959, 1960).

Pronator teres (LSB 83.4 g; Figs. 27–29)
- Usual attachments: From the humerus (caput humerale) and usually also from the ulna (caput ulnare), to the radius.
- Usual innervation: Median nerve, which usually passes between the ulnar and humeral heads of the muscle (Champneys 1872, Chapman 1879, Dwight 1885, Hepburn 1892, Sonntag 1923, Swindler & Wood 1973: *P. troglodytes*; Miller 1952: *P. paniscus*).
- Notes: Gratiolet & Alix (1866) described a humeral head only in *Pan*, and Lewis (1989) stated that a distinct ulnar head is not present in apes. However, the chimpanzees reported by Tyson (1699), Macalister (1871), Champneys (1872), Chapman (1879), Hepburn (1892), Beddard (1893), Dwight (1895), Sonntag (1923, 1924), Miller (1952), Swindler & Wood (1973), Stern & Larson (2001) and dissected by us have an ulnar head and a humeral head separated by the median nerve, and Oishi et al. (2009) found two heads in two of the four *Pan* specimens dissected by them, but they could not find an ulnar head in the other two. According to Parsons (1898b), Loth (1931) and Jouffroy (1971) two heads are present in 90%, 91% and 90% of chimpanzees, respectively, but Gibbs (1999) stated that two heads are found in 5 out of 9 *Pan* specimens.
- Synonymy: Grand pronateur or rond pronateur (Gratiolet & Alix 1866); pronator radii teres (Champneys 1872, Hepburn 1892, Beddard 1893, Sonntag 1923).

Palmaris brevis (LSB 1.4 g; Fig. 39)
- Usual attachments: From the pisiform and/or the flexor retinaculum to the skin of the medial border of the palm.

- Usual innervation: Superficial branch of ulnar nerve (Miller 1952: *P. paniscus*; Swindler & Wood 1973: *P. troglodytes*).
- Notes: The palmaris brevis was not present in the chimpanzee specimen dissected by Hepburn (1892), or in the two specimens of Sarmiento (1994) and in the specimen of Wilder (1862), but it was present in the specimen of Vrolik (1841), the specimen of Gratiolet & Alix (1866), the two specimens of Humphry (1867), the specimen of Champneys (1872), the specimen of Sonntag (1923), the specimen of Miller (1952), the specimen illustrated by Dylevsky (1967), the specimen of Swindler & Wood (1973) and in 4 of the 5 dissected specimens in which we could discern this feature. According to the reviews of the literature by Sarmiento (1994) and Gibbs (1999) this muscle is present in 7/9 and 4/5 *Pan*, respectively, and according to our own review of the literature and the data obtained from our dissections, it is present in 13/18 chimpanzees. The **flexor digitorum brevis manus** and **palmaris superficialis** (e.g., Diogo et al. 2009a) are usually not present as distinct muscles in chimpanzees.
- Synonymy: Palmaire cutané (Gratiolet & Alix 1866).

Lumbricales (LSB lumbricalis 1 = 3.0 g; LSB lumbricalis 2 = 5.7 g; LSB lumbricalis 3 = 3.8 g; LSB lumbricalis 4 = 1.1 g; Figs. 28, 37, 39–40)
- Usual attachments: From the dorsal surfaces of the tendons of the flexor digitorum profundus to digits 2, 3, 4 and 5, to the radial side of the proximal phalanx and the extensor expansion of digits 2 (lumbricalis I), 3 (lumbricalis II), 4 (lumbricalis III) and 5 (lumbricalis IV).
- Usual innervation: First, second and third lumbricales by median nerve, and the fourth by the deep branch of ulnar nerve (Hepburn 1892: *P. troglodytes*); first and second lumbricales by the median nerve and the third and fourth lumbricales by the deep branch of the ulnar nerve (Miller 1952: *P. paniscus*; Swindler & Wood 1973: *P. troglodytes*).
- Function: See flexores breves profundi, below.
- Notes: The **intercapitulares** (see, e.g., Jouffroy 1971) are usually not present in chimpanzees.

Contrahentes digitorum
- Usual attachments: Mainly from the contrahens fascia to the base of proximal phalanx and to the extensor expansions of the ulnar side of digits 4 (contrahens I) and 5 (contrahens II), although sometimes there are contrahentes to other digits (e.g., to digit 2: Brooks 1886a).
- Usual innervation: Deep branch of the ulnar nerve (Hepburn 1892: *P. troglodytes*; Miller 1952: *P. paniscus*).
- Notes: In the chimpanzee specimen dissected by Gratiolet & Alix (1866), the two specimens dissected by Jouffroy & Lessertisseur (1959), the two specimens reported by Day & Napier (1963) and four of the five specimens dissected by us in which we could discern this feature the contrahentes were not present

as fleshy, well-defined, distinct structures (the exception being PFA 1077). In the specimen dissected by Brooks (1886a) there were three contrahentes to digits 2, 4 and 5. However, according to Lewis (1989) in *Pan* there are usually contrahentes to digits 4 and 5; this was the case in a specimen described by Hartmann (1886), in the specimen dissected by Hepburn (1892), the specimen dissected by Sarmiento (1994), the specimen dissected by Miller (1952), and the specimen reported by Swindler & Wood (1973). In the review of the literature by Sarmiento (1994) the contrahentes were said to be present in 7 out of 7 *Pan*.
- Synonymy: Part of the adductors of digits (Brooks 1886a); contrahentes manus (Swindler & Wood 1973).

Adductor pollicis (LSB 13.9 g; Figs. 32–33, 35, 37–40)
- Usual attachments: The caput obliquum and caput transversum are usually well-differentiated, connecting the metacarpals II and/or III, the contrahens fascia, and often at least some carpal bones and/or ligaments, to the metacarpophalangeal joint, as well as to the proximal phalanx and sometimes also to the distal phalanx of the thumb (directly by means of a thin tendon and/or indirectly by means of a ligament) and occasionally to metacarpal I (e.g., Gratiolet & Alix 1866, Jouffroy & Lessertisseur 1960). When present, the adductor pollicis accessorius (see below) often connects the proximo-medial portion of the metacarpal I and/or adjacent carpal structures to the metacarpophalangeal joint and/or the proximal portion of the proximal phalanx of digit 1.
- Usual innervation: Deep branch of the ulnar nerve (Brooks 1887, Hepburn 1892: *P. troglodytes*; Miller 1952: *P. paniscus*); ulnar nerve (Swindler & Wood 1973: *P. troglodytes*).
- Notes: There has been much controversy regarding the homologies of the thenar muscles of primate and non-primate mammals. This subject has been discussed in detail in the recent studies of Diogo et al. (2009a) and particularly of Diogo & Abdala (2010) and Diogo & Wood (2011) and here we summarize the main conclusions of those studies. The **'interosseous volaris primus of Henle'** of modern human anatomy corresponds to a **thin, deep additional slip of the adductor pollicis** (**TDAS-AD** *sensu* Diogo & Abdala 2010 and Diogo & Wood 2011; **adductor pollicis accessorius** *sensu* Diogo et al. in press and *sensu* the present study), and not to the **flexor brevis profundus 2** of 'lower' mammals, as suggested by some authors and as suggested by the use of the name 'interosseous volaris primus of Henle'). Within chimpanzees, the adductor pollicis accessorius is present in some, but not in most, cases (i.e., in < 50% of the cases). It was present in the specimen dissected by Champneys (1872) and was probably present in the specimen dissected by Gratiolet & Alix (1866) and the specimen reported by Brooks (1886a), and in 25% of chimpanzees

according to Tuttle (1970), in 2 out of 7 chimpanzees according to the review of the literature by Sarmiento (1994), and in at least one of the hands of the 4 specimens dissected by us in which we could discern this feature in detail. It was not found in the two specimens dissected by Susman et al. (1999) and in the two specimens dissected by Sarmiento (1994). Abramowitz (1955) also supports the idea that the adductor pollicis accessorius is present in at least some chimpanzees. The adductor pollicis accessorius is also present in some, but not in most (i.e., in < 50% of the cases) gorillas, and it is present in more than 50% of modern humans (Diogo & Wood 2011, 2012). Regarding the **'deep head of the flexor pollicis brevis'** of modern human anatomy, this corresponds to the flexor brevis profundus 2 of 'lower' mammals. This structure is clearly found in four of the five chimpanzees dissected by us in which we could discern this feature in detail (including the right hand of one of the infants, PFA 1077; we could not find it in the left hand of this infant nor in the single—left—dissected hand of the infant PFA UNC) and in most of the specimens dissected by other authors, who often refer to this structure as the 'deep head of the flexor pollicis brevis'. Regarding the **'superficial head of the flexor pollicis brevis'** of modern human anatomy, this thus seems to correspond to a true **flexor pollicis brevis**. The true flexor pollicis brevis and the **opponens pollicis** likely derive from the **flexor brevis profundus 1** of 'lower' mammals, and they are present as distinct muscles in the vast majority of chimpanzees (see below).
- Synonymy: Adductor pollicis and probably the 'ulnar division of flexor pollicis brevis' (Humphry 1867); adducteur du pouce (Gratiolet & Alix 1866); adductor obliquus pollicis plus adductor transversus pollicis (Sonntag 1923); contrahentes I (Jouffroy 1971).

Flexores breves profundi (LSB flexor brevis profundus 4, or 'interosseous palmaris 1' = 8.6 g; LSB flexor brevis profundus 7, or 'interosseous palmaris 2' = 6.4 g; LSB flexor brevis profundus 7, or 'interosseous palmaris 3' = 5.0 g; Figs. 32, 34–36, 40)
- Usual attachments: Flexores breves profundi 3 and 4 run mainly from metacarpal II to the radial and ulnar sides of the proximal phalanx of digit 2, respectively; flexores breves profundi 5 and 6 run mainly from metacarpal III to the radial and ulnar sides of the proximal phalanx of digit 3, respectively; flexores breves profundi 7 and 8 run mainly from metacarpal IV to the radial and ulnar sides of the proximal phalanx of digit 4, respectively; flexor brevis profundus 9 runs mainly from metacarpal V to the ulnar side of the proximal phalanx of digit 5.
- Usual innervation: Deep branch of the ulnar nerve (Hepburn 1892: *P. troglodytes*; Miller 1952: *P. paniscus*).
- Function: EMG study of the lumbricales and of the 'palmar and dorsal interossei' (flexores breves profundi plus intermetacarpales *sensu* the present study) of

chimpanzees indicates that the function of these muscles is essentially similar to that of modern humans, the 'interossei' being mainly associated with flexion of the metacarpophalangeal joints and extension of the interphalangeal joints, as in modern humans (could not be accessed if the 'interossei' of chimpanzees were also associated with abduction/adduction and/or rotation of the digits) and the lumbricales with extension of the interphalangeal joints, as in modern humans (Susman & Stern 1980).

- Notes: As explained above (see adductor pollicis and interossei palmares), the **flexor brevis profundus 2** of 'lower' mammals corresponds very likely to the **'deep head of the flexor pollicis brevis'** of modern human anatomy; the **'superficial head of the flexor pollicis brevis'** of modern human anatomy, as well as the **opponens pollicis**, derive very likely from the **flexor brevis profundus 1** of 'lower' mammals, while the **flexor digiti minimi brevis** and the **opponens digiti minimi** derive very likely from the **flexor brevis profundus 10** of 'lower' mammals (for recent reviews, see Diogo et al. 2009a, Diogo & Abdala 2010 and Diogo & Wood 2012). In extant hominoids except *Pan* the **flexores breves profundi** 3, 5, 6 and 8 are fused with the **intermetacarpales** 1, 2, 3 and 4, forming the interossei dorsales 1, 2, 3 and 4, respectively; the **interossei palmares** 1, 2 and 3 of extant hominoids except *Pan* thus correspond respectively to the flexores breves profundi 4, 7 and 9 of 'lower' mammals and of primates such as chimpanzees (e.g., Lewis 1989, Diogo et al. 2009a, and Diogo & Abdala 2010). The **interossei accessorii** are not present as distinct muscles in chimpanzees.
- Synonymy: Palmar or volar interossei and part of dorsal interossei (Huxley 1864, Champneys 1872, Dwight 1895, Miller 1952, Tuttle 1969, 1970, Swindler & Wood 1973, Gibbs 1999, Ogihara et al. 2005, Tocheri et al. 2008); palmar interossei plus interosseous volaris primus (Brooks 1887; Hepburn 1892); palmar interossei and possibly part of flexor brevis pollicis (Sonntag 1923); flexores breves (Lewis 1989).

Intermetacarpales (LSB intermetacarpalis 1 + flexor brevis profundus 3, or 'interosseous dorsalis' 1 = 21.0 g; LSB intermetacarpalis 2 + flexor brevis profundus 5, or 'interosseous dorsalis' 2 = 14.5 g; LSB intermetacarpalis 3 + flexor brevis profundus 6, or 'interosseous dorsalis' 3 = 9.9 g; LSB intermetacarpalis 4 + flexor brevis profundus 8, or 'interosseous dorsalis' 4 = 10.2 g; Figs. 30, 34, 36)
- Usual attachments: Intermetacarpalis 1 runs mainly from metacarpals I and II to the radial side of the proximal phalanx and extensor expansion of digit 2; intermetacarpalis 2 mainly from metacarpals II and III to the radial side of the proximal phalanx and extensor expansion of digit 3; intermetacarpalis 3 mainly from metacarpals III and IV to the ulnar side of the proximal phalanx and extensor expansion of digit 3; intermetacarpalis 4 mainly from metacarpals IV and V to the ulnar side of the proximal phalanx and extensor expansion of digit 4.

- Usual innervation: Deep branch of the ulnar nerve (Hepburn 1892: *P. troglodytes*; Miller 1952: *P. paniscus*); ulnar nerve (Swindler & Wood 1973: *P. troglodytes*).
- Notes: The **interdigitales** present in primates such as lorisoids are usually not present in chimpanzees.
- Synonymy: Part or totality of dorsal interossei (Gratiolet & Alix 1866, Champneys 1872, Brooks 1886a, Hepburn 1892, Sonntag 1923, 1924, Miller 1952, Tuttle 1969, 1970, Swindler & Wood 1973, Susman & Stern 1980, Lewis 1989, Gibbs 1999, Ogihara et al. 2005, Tocheri et al. 2008).

Flexor pollicis brevis and ***flexor brevis profundus 2*** (LSB flexor pollicis brevis + flexor brevis profundus 2, or 'superficial + deep heads of flexor pollicis brevis' = 7.8 g; Figs. 32, 39–40)
- Usual attachments: As explained above (see adductor pollicis), the flexor brevis profundus 2 of 'lower' mammals corresponds to the **'deep head of the flexor pollicis brevis'** of modern human anatomy; the **'superficial head of the flexor pollicis brevis'** of modern human anatomy thus corresponds to a true flexor pollicis brevis, being likely derived, together with the **opponens pollicis**, from the **flexor brevis profundus 1** of 'lower' mammals (for recent reviews see Diogo et al. 2009a, Diogo & Abdala 2010 and Diogo & Wood 2012). The flexor brevis profundus 2 of chimpanzees usually runs mainly from the carpal region to the ulnar side of the proximal phalanx of digit 1, while the true flexor pollicis brevis runs mainly from the carpal region to the radial side of the proximal phalanx of digit 1, lying superficially to the flexor brevis profundus 2 (see Notes below).
- Usual innervation of the true flexor pollicis brevis and of the flexor brevis profundus 2: 'flexor pollicis brevis' (which might include the flexor pollicis brevis plus flexor brevis profundus 2 *sensu* the present study) is innervated by the median nerve (Champneys 1872: *P. troglodytes*); 'superficial head of the flexor pollicis brevis' of modern human anatomy (i.e., flexor pollicis brevis *sensu* the present study) is innervated by the median nerve (Brooks 1887, Hepburn 1892: *P. troglodytes*); 'flexor pollicis brevis' (i.e., flexor pollicis brevis plus flexor brevis profundus 2 *sensu* the present study) is innervated by the ulnar and the median nerves (Miller 1952: *P. paniscus*); median nerve ('superficial head', because the 'deep head' of modern human anatomy is, according to these authors, absent in *Pan*; Swindler & Wood 1973: *P. troglodytes*); the true flexor pollicis brevis is innervated by the median nerve, while the flexor brevis profundus 2 innervated by the ulnar nerve (our specimens GWUANT PT1 and GWUANT PT2).
- Notes: As there is much confusion in the literature about the flexor brevis profundus 2 and the true flexor pollicis brevis *sensu* the present study, we will describe in some detail the configuration found in some of the chimpanzees dissected by us. For more details about the descriptions of other authors and for information about the synonymy, homologies and attachments of these

muscles, see Diogo & Wood (2012) and see also the sections above about adductor pollicis, interossei palmares and intermetacarpales. In our adult specimens GWUANT PT1 and GWUANT PT2 the true flexor pollicis brevis originates from the flexor retinaculum and trapezium, while the flexor brevis profundus 2 originates from the capitate and trapezoid. The true flexor pollicis brevis inserts onto the lateral side of the palmar surface of the base of the proximal phalanx of digit 1 and/or on the adjacent sesamoid bone, while the flexor brevis profundus 2 inserts onto the medial side of the palmar surface of the proximal phalanx of digit 1 and/or on the adjacent sesamoid bone. On the left side of our infant specimen PFA 1077 only one 'head' of the so-called 'flexor pollicis brevis' seems to be present (i.e., the flexor brevis profundus 2 does not seem to be present as a separate muscle). The only 'head' that is present is the 'superficial head' (i.e., the true flexor pollicis brevis *sensu* the present study, that runs from the flexor retinaculum, the trapezium and the ligaments of the carpal region to the central and lateral portions of the ventral region of the metacarpophalangeal joint of the thumb). However, in the right hand of this infant specimen there is what clearly seems to be a 'deep head of the flexor pollicis brevis'; the flexor brevis profundus 2 does seem to be present as a distinct muscle, being somewhat blended with the opponens pollicis. On the left dissected hand of our infant specimen PFA UNC the 'deep head of the flexor pollicis brevis' (i.e., the flexor brevis profundus 2) is not present as a distinct muscle, for the single 'head of the flexor pollicis brevis' (i.e., the true flexor pollicis brevis *sensu* the present study) is blended dorsally to the opponens pollicis.
- Synonymy of the true flexor pollicis brevis: Part or totality of court fléchisseur du pouce (Gratiolet & Alix 1866); part or totality of flexor brevis pollicis, and probably of the 'interosseous volaris primus' (Champneys 1872); part or totality of flexor brevis pollicis and possibly part of the abductor pollicis brevis (Sonntag 1923); 'superficial head of the flexor pollicis brevis' of human anatomy (Terminologia Anatomica 1998).

Opponens pollicis (LSB 5.9 g; Figs. 33, 39)
- Usual attachments: From the flexor retinaculum, trapezium and/or adjacent sesamoid bone to the whole length of metacarpal I.
- Usual innervation: Median nerve (Brooks 1887, Hepburn 1892, Swindler & Wood 1973: *P. troglodytes*; Miller 1952: *P. paniscus*).
- Notes: Opposant du pouce (Gratiolet & Alix 1866).

Flexor digiti minimi brevis (LSB 5.2 g; Figs. 31, 35, 39)
- Usual attachments: From the hamate and/or the flexor retinaculum, to the ulnar side of the proximal phalanx.
- Usual innervation: Ulnar nerve (Hepburn 1892, Swindler & Wood 1973: *P. troglodytes*; Miller 1952: *P. paniscus*).

- Notes: In *Pan* an origin from the hamate and flexor retinaculum was found by Gratiolet & Alix (1866), Sonntag (1923) and Swindler & Wood (1973), while Miller (1952) decribed an origin from the flexor retinaculum only and we found an origin from the flexor retinaculum and, often, also from the hamate in our specimens.
- Synonymy: Court fléchisseur du cinquième doigt (Gratiolet & Alix 1866); flexor brevis minimi digiti (Champneys 1872, Sonntag 1923); flexor digiti quinti (Miller 1952).

Opponens digiti minimi (LSB 6.7 g; Figs. 31, 35, 39–40)
- Usual attachments: From the hamate and, often, also from the flexor retinaculum to the whole length of metacarpal V, being often divided into a superficial and a deep head (see Notes below); occasionally the muscle may also originate for structures such as the contrahens aponeurosis (e.g., Gratiolet & Alix 1866) and/or the transverse carpal ligament (e.g., Miller 1952).
- Usual innervation: Ulnar nerve (Hepburn 1892, Swindler & Wood 1973: *P. troglodytes*; Miller 1952: *P. paniscus*).
- Notes: As noted by Brooks (1886a), Lewis (1989) and Diogo et al. (2009a) and Diogo & Wood (2011, 2012) and corroborated by our dissections, in hominoids the opponens digiti minimi is usually slightly differentiated into superficial and deep bundles. Lewis (1989) stated that the opponens digiti minimi is more markedly divided in hominoids such as *Pan* and modern humans than in hominoids such as hylobatids, and Brooks (1886a) stated that, contrary to *Pan* and modern humans, in hominoids such as *Pongo* there are no superficial and deep bundles of the muscle separated by the deep branch of the ulnar nerve. Regarding our dissections, in hylobatids, *Gorilla* and *Pongo* the deep branch of the ulnar nerve runs mainly radial to both these bundles, and not mainly superficially (palmar) to the deep bundle and deep (dorsal) to the superficial bundle as is usually the case in *Pan* and particularly in modern humans.
- Synonymy: Opposant du cinquième doigt (Gratiolet & Alix 1866); opponens minimi digiti (Champneys 1872, Sonntag 1923); opponens digiti quinti (Miller 1952).

Abductor pollicis brevis (LSB 5.7 g; Figs. 32, 35, 38–39)
- Usual attachments: From the flexor retinaculum and sometimes from structures such as the trapezium, the adjacent sesamoid bone, the transverse carpal ligament and/or the scaphoid, to the radial side of the metacarpophalangeal joint, its sesamoid bone and/or the base of the proximal phalanx of digit 1; occasionally the insertion of the muscle extends to the distal phalanx of this digit (e.g., Hepburn 1892, Dwight 1895) and/or to the distal portion of metacarpal I (e.g., Hepburn 1892, Aziz & Dunlap 1986, Landsmeer 1986).
- Usual innervation: Median nerve (Brooks 1887, Hepburn 1892, Swindler & Wood 1973: *P. troglodytes*; Miller 1952: *P. paniscus*).

- Notes: In chimpanzees the abductor pollicis brevis may be divided into slips (e.g., Aziz & Dunlap 1986) and may be reinforced by slips from the 'flexor pollicis brevis' (true flexor pollicis brevis plus flexor brevis profundus 2 *sensu* the present study) according to Hepburn (1892).
- Synonymy: Court abducteur du pouce (Gratiolet & Alix 1866); abductor pollicis (Champneys 1872); part or totality of abductor pollicis brevis (Sonntag 1923).

Abductor digiti minimi (LSB 13.5 g; Figs. 31, 35, 39)
- Usual attachments: From the pisiform and sometimes from the flexor retinaculum (e.g., Swindler & Wood 1973; some of our specimens) and/or hamate (e.g., Swindler & Wood 1973; some of our specimens), to the ulnar side of the proximal phalanx of digit 5 and sometimes to the adjacent metacarpophalangeal joint (e.g., some of our specimens), sesamoid bone (e.g., some of our specimens), and/or the dorsal extensor aponeurosis (e.g., Miller 1952; some of our specimens).
- Usual innervation: Radial nerve (Hepburn 1892, Sonntag 1923, Swindler & Wood 1973: *P. troglodytes*; Miller 1952; *P. paniscus*).
- Synonymy: Métacarpien du index or métacarpien du deuxième doigt or deuxiéme métacarpien dorsal or prémier radial externe (Gratiolet & Alix 1866); extensor carpi radialis longior (Champneys 1872, Hepburn 1892, Beddard 1893, Sonntag 1923).

Extensor carpi radialis longus (LSB 48.8 g; Fig. 30)
- Usual attachments: From the lateral supracondylar ridge and often from the lateral epicondyle of the humerus, to the base of metacarpal II and often also to base of metacarpal I (e.g., Bojsen-Møller 1978; our specimen HU PT1 and one side of our specimen PFA 1077); in a few cases it also attaches to other structures such as the base of metacarpal III (e.g., Payne 2001) and/or the ligament connecting the metacarpal II to the trapezium (e.g., Gratiolet & Alix 1866).
- Usual innervation: Radial nerve (Hepburn 1892, Sonntag 1923, Swindler & Wood 1973: *P. troglodytes*; Miller 1952: *P. paniscus*).
- Synonymy: Premier radial (Deniker 1885); extensor carpi radialis longior (Hartmann 1886, Hepburn 1892).

Extensor carpi radialis brevis (LSB 61.6 g; Fig. 30)
- Usual attachments: From the lateral epicondyle of the humerus and occasionally also from structures such as the radial collateral ligament (e.g., Sonntag 1923), to the base of metacarpal III and often also to the base of metacarpal II (e.g., Ribbing & Hermansson 1912; our specimens HU PT1, PFA 1077 and PFA UNC).
- Usual innervation: Posterior interosseous nerve (Hepburn 1892: *P. troglodytes*); radial nerve (Swindler & Wood 1973: *P. troglodytes*; Miller 1952: *P. paniscus*).

- Synonymy: Métacarpien du médius or métacarpien du troisième doigt or troisième métacarpein dorsal or deuxième radial externe (Gratiolet & Alix 1866); extensor carpi radialis brevior (Champneys 1872, Hepburn 1892, Beddard 1893, Sonntag 1923).

Brachioradialis (LSB 194.1 g; Figs. 27–30)
- Usual attachments: From the shaft and distal end of the humerus to the shaft of the radius (usually not reaching the styloid process; exceptions were reported by, e.g., Hepburn 1892, Ziegler 1964 and Swindler & Wood 1973 and found in our specimens PFA 1077 and PFA UNC). The origin of the brachioradialis extends more proximally in *Pan* than in modern humans (e.g., often lying at the level of the deltoid insertion: Sonntag 1923 and Swindler & Wood 1973 and our specimens GWUANT PP1, GWUANT PP2, PFA 1016, PFA 1051 and HU PT1).
- Usual innervation: Radial nerve (Hepburn 1892, Sonntag 1923, Straus 1941a,b, Swindler & Wood 1973: *P. troglodytes*; Miller 1952: *P. paniscus*).
- Synonymy: Supinator longus (Wyman 1855; Wilder 1862; Gratiolet & Alix 1866, Macalister 1871, Barnard 1875, Duckworth 1904, Sonntag 1923); supinator radii longus (Champneys 1872, Hepburn 1892, Beddard 1893).

Supinator (LSB 122.2 g; Figs. 28, 30)
- Usual attachments: From the lateral epicondyle of the humerus (caput humerale, or superficiale) and usually also from the proximal part of the ulna (caput ulnare, or profundum; e.g., Gratiolet & Alix 1866, Champneys 1872, Beddard 1893, Straus 1941a, Miller 1952, Swindler & Wood 1973 and our dissections), to the proximal radius.
- Usual innervation: Posterior interosseous (deep radial) nerve (Champneys 1872, Hepburn 1892, Sonntag 1923, Straus 1941a,b: *P. troglodytes*; Miller 1952: *P. paniscus*); radial nerve (Swindler & Wood 1973: *P. troglodytes*).
- Notes: The supinator is usually perforated by the radial nerve in chimpanzees (e.g., Macalister 187, Beddard 1893, Straus 1941a, Lewis 1989 and our specimens PFA 1077 and PFA UNC).
- Synonymy: Court supinateur (Gratiolet & Alix 1866); supinator radii brevis (Macalister 1871, Champneys 1872, Hepburn 1892, Beddard 1893); supinator brevis (Sonntag 1923).

Extensor carpi ulnaris (LSB 44.8 g; Figs. 29–30)
- Usual attachments: From the lateral epicondyle of the humerus (caput humerale) and usually also the ulna (caput ulnare; e.g., Gratiolet & Alix 1866, Champneys 1872, Sonntag 1923, Straus 1941a, Miller 1952, Swindler & Wood 1973 and our dissections) to the base of metacarpal V and occasionally to the proximal phalanx of digit 5 (e.g., Macalister 1871: see Notes below).

- Usual innervation: Posterior interosseous (deep radial) nerve (Hepburn 1892: *P. troglodytes*; Miller 1952: *P. paniscus*); radial nerve (Swindler & Wood 1973: *P. troglodytes*).
- Notes: Macalister (1871) stated that in the *P. troglodytes* specimen dissected by him the extensor carpi ulnaris had a distinct **'ulnaris-quinti'** tendon to the proximal phalanx of digit 5.
- Synonymy: Métacarpien du cinquième doigt or cubital postérieur or cubital dorsal or cinquième métacarpien dorsal (Gratiolet & Alix 1866).

Anconeus (LSB 9.0 g; Fig. 30)
- Usual attachments: From the lateral epicondyle of the humerus to the olecranon process of the ulna and in some cases also to adjacent areas of the ulna (e.g., Miller 1952, Swindler & wood 1973, Payne 2001, and our dissections).
- Usual innervation: Radial nerve (Swindler & Wood: *P. troglodytes*; Miller 1952: *P. paniscus*); ulnar nerve (Ziegler 1964: *P. troglodytes*).
- Notes: The anconeus is usually present as a distinct muscle in *Pan* according to Gratiolet & Alix (1866), Macalister (1871), Champneys (1872), Sonntag (1923), Miller (1952), Swindler & Wood (1973), Payne (2001) and to our dissections.

Extensor digitorum (LSB 128.1 g; Figs. 29–30, 34)
- Usual attachments: From the lateral epicondyle of the humerus and occasionaly also from the radius (e.g., MacDowell 1910) and/or the ulna (e.g., MacDowell 1910 and our specimens PFA 1077 and PFA UNC), to the middle phalanges and (via the extensor expansions) to the distal phalanges of digits 2, 3, 4 and 5; occasionally the muscle has an insertion onto the interphalangeal joints (e.g., Sonntag 1923) or has no insertion onto digit 5 (e.g., Vrolik 1841, Wilder 1862 and Straus 1941a).
- Usual innervation: Posterior interosseous (deep radial) nerve (Hepburn 1892: *P. troglodytes*; Miller 1952: *P. paniscus*); radial nerve (Swindler & Wood 1973: *P. troglodytes*).
- Synonymy: Extensor communis digitorum (Wilder 1862, Champneys 1872, Hepburn 1892, Beddard 1893, Sonntag 1923); extenseurs directes des doigts (Gratiolet & Alix 1866); extensor digitorum longus (Macalister 1871); extensor digitorum communis (Barnard 1875, Dwight 1895, MacDowell 1910, Straus 1941a,b, Miller 1952).

Extensor digiti minimi (LSB 24.3 g; Figs. 29–30)
- Usual attachments: From the lateral epicondyle of the humerus and/or the common extensor tendon, and/or occasionally from the ulna (e.g., MacDowell 1910), to the middle phalanx and (via the extensor expansion) to the distal phalanx of digit 5; the muscle often also inserts onto digit 4 (e.g., Fig. 29; see Notes below).

- Usual innervation: Posterior interosseous (deep radial) nerve (Hepburn 1892: *P. troglodytes*; Miller 1952: *P. paniscus*); radial nerve (Swindler & Wood 1973: *P. troglodytes*).
- Notes: In the *Pan* specimens described by Wilder (1862), Gratiolet & Alix (1866), Macalister (1871), Champneys (1872), Chapman (1879), Hepburn (1892), MacDowell (1910), Sonntag (1923), Miller (1952), Swindler & Wood (1973), Aziz & Dunlap (1986), Landsmeer (1986), and Oishi et al. (2009), as well as in 3 of the 6 specimens dissected by us in which we could discern this feature in detail, as well as in three specimens dissected by Straus (1941a) and in one specimen dissected by Kaneff (1980a), the muscle goes to digit 5 only. In the 3 other specimens dissected by us in which we could discern this feature in detail, in one specimen dissected by Huxley (1864), in one specimen dissected by Dwight (1895), in one specimen dissected by Straus (1941a), in 4 specimens dissected by Kaneff (1980a), and in one specimen dissected by Lewis (1989), the muscle goes to digits 4 and 5. According to the review of the literature by Gibbs (1999) an exclusive insertion onto digit 5 occurs in 26 out of 31 (84%) of *Pan*; it occurs in 73% *Pan* according to the review of the literature done by Straus (1941a). The **extensor digiti quarti** is usually not present as a distinct muscle in chimpanzees.
- Synonymy: Extensor minimi digiti (Wilder 1862, Macalister 1871, Champneys 1872, Hepburn 1892, Beddard 1893, Dwight 1895, Sonntag 1923); extenseur latéral du cinquième doigt (Gratiolet & Alix 1866); extensor digiti quinti proprius (MacDowell 1910); extensor digiti—quarti et—quinti proprius (Straus 1941a,b); extensor digiti quinti proprius (Miller 1952); extensor digitorum lateralis bidigitalis (Kaneff 1979, 1980a); extensor digitorum proprius or profundus 4 and 5 (Lewis 1989).

Extensor indicis (LSB 9.3 g; Figs. 29–30)
- Usual attachments: From the ulna and often also from the interosseous membrane and occasionally also from the radius (e.g., Sonntag 1923) and/or the humerus/common extensor tendon (e.g., Gratiolet & Alix 1866), to digit 2; occasionally there is also an insertion onto digit 4 and/or digit 3 and/or no insertion onto digit 2 (see Notes below).
- Usual innervation: Posterior interosseous (deep radial) nerve (Hepburn 1892: *P. troglodytes*; Miller 1952: *P. paniscus*); radial nerve (Swindler & Wood 1973: *P. troglodytes*).
- Notes: An insertion onto digit 2 only is described in *Pan* by most authors, including Wilder (1862), Gratiolet & Alix (1866), Broca(1869), Champneys (1872), Chapman (1879), Beddard (1893), Dwight (1895), Sonntag (1923), Fick (1925), Miller (1952), Tuttle (1969), Swindler & Wood (1973), Dunlap et al. (1985), Landsmeer (1986), Lewis (1989) and Oishi et al. (2009) and found in the vast majority of the specimens dissected by us. However, Hepburn (1892)

found an insertion onto digits 2 and 4, Straus (1941a) and Kaneff (1980ab) onto digit 2 or onto digits 2 and 3, and Humphry (1867), Macalister (1871), Hartmann (1886), MacDowell (1910), Ribbing & Hermansson (1912) and Jouffroy & Lessertisseur (1957) onto digits 2 and 3. According to the review of the literature by Straus (1941ab), in *Pan* an insertion onto digits 2, 3 and 4 occurs in about 4% of the cases, onto digits 2 and 3 occurs in about 21% of the cases, onto digits 2 and 4 occurs in about 4% of the cases, only onto digit 3 occurs in about 4% of the cases, and only onto digit 2 occurs in about 68% of the cases. In our infant specimen PFA 1077 the extensor indicis has tendons to digits 2 and 3, but the tendon to digit 3 is very 'degenerated' distally; this suggests that the tendon to digit 3 could become completely degenerated/lost during ontogeny, thus leading to an adult configuration in which there is only a tendon to digit 2, as is usually the case in adult chimps. The **extensor digiti III proprius** and the **extensor brevis digitorum manus** (e.g., Diogo et al. 2009a) are usually not present as distinct muscles in chimpanzees.
- Synonymy: Extenseur latéral de l'index (Gratiolet & Alix 1866); extenseur propre de l'index (Broca 1869); indicator (Macalister 1871); part or totality of extensor profundus digitorum or extensor digitorum profundus (Barnard 1875, Hepburn 1892, Straus 1941a,b); extensor indicis proprius (Dwight 1895, MacDowell 1910, Miller 1952); extensor digiti indicis (Swindler & Wood 1973); extensor digitorum profundus proprius or extensor indicis-et-medii digiti (Aziz & Dunlap 1986); extensor digitorum proprius or profundus 2 (Lewis 1989).

Extensor pollicis longus (LSB 31.3 g; Figs. 29–30)
- Usual attachments: From the ulna and interosseous membrane and occasionally from the humerus/common extensor tendon (e.g., Gratiolet & alix 1866) and/or the radius (e.g., some of our specimens), to the distal phalanx of the thumb and sometimes also to its proximal phalanx (e.g., Champneys 1872, Sutton 1883, Ziegler 1964 and our specimen PFA UNC); occasionally there is also an insertion onto digit 2 (see Notes below).
- Usual innervation: Posterior interosseous (deep radial) nerve (Hepburn 1892: *P. troglodytes*; Miller 1952: *P. paniscus*).
- Notes: The structure designated as **'extensor communis pollicis et indicis'** in one of the five chimpanzees dissected by Kaneff (1980a) does not correspond to the extensor communis pollicis et indicis *sensu* the present study. Instead, it corresponds to the extensor pollicis longus *sensu* the present study which in this specimen goes to both digits 1 and 2. In the four other chimpanzees dissected by Kaneff 1980a the extensor pollicis longus goes to digit 1 only, as is usually the case in most chimpanzees.
- Synonymy: Extenseur latéral du pouce (Gratiolet & Alix 1866); extensor secundi internodii pollicis (Champneys 1872, Sutton 1883, Hepburn 1892, Beddard

1893, Sonntag 1923); part or totality of extensor profundus digitorum or extensor digitorum profundus (Straus 1941a,b); extensor digitorum proprius or profundus 1 (Lewis 1989).

Abductor pollicis longus (LSB 38.7 g; Figs. 29–30)
- Usual attachments: From the interosseous membrane and/or ulna and/or radius, to metacarpal I and trapezium; sometimes the muscle inserts also/instead onto the scaphoid and/or sesamoid bone adjacent to the trapezium, and ocasionally the muscle extends to the proximal phalanx of digit 1 (see Notes below).
- Usual innervation: Posterior interosseous (deep radial) nerve (Hepburn 1892: *P. troglodytes*; Miller 1952: *P. paniscus*); radial nerve (Swindler & Wood 1973: *P. troglodytes*).
- Notes: Apart from modern humans, in all of the primate specimens dissected by us only in hylobatids is there is a distinct **extensor pollicis brevis** that is only partially blended, proximally, with the belly of the abductor pollicis longus (e.g., Diogo & Wood 2011, 2012). It should be noted that some authors have described an 'extensor pollicis brevis' and an 'abductor pollicis longus' in primate taxa other than hylobatids and modern humans, including chimpanzees (see Synonymy below). For example, in gorillas the name 'extensor pollicis brevis' has been used (e.g., Hepburn 1892, Straus 1941ab) to refer to a tendon of the abductor pollicis longus (*sensu* the present study) that inserts onto the proximal phalanx of the thumb (i.e., to the typical insertion point of the extensor pollicis brevis of modern humans). However, as stressed by authors such as Kaneff (1979, 1980a,b) and Aziz & Dunlap (1986) and corroborated by our dissections in gorillas there is usually a single fleshy belly of the abductor pollicis longus that then gives rise to the so-called 'tendons of the extensor pollicis brevis and of the abductor pollicis longus'; this configuration is usually also the case in *Pongo* and *Pan*. Thus, contrary to the condition in *Homo* and hylobatids, in *Pongo*, *Pan* and *Gorilla* the extensor pollicis brevis is usually *not* present as a separate muscle. In *Pan* the insertion of the abductor pollicis longus is onto the metacarpal I, the adjacent sesamoid bone and trapezium according to Hepburn (1892), Dwight (1895) and Sonntag (1923), to the scaphoid and metacarpal I according to Humphry (1867), to the sesamoid, scaphoid and metacarpal I according to Ziegler (1964), and to the metacarpal I and trapezium according to Vrolik (1841), Wyman (1855), Wilder (1862), Huxley (1864), Champneys (1872), Hartmann (1886), Beddard (1893), Sutton (1883), Gratiolet & Alix (1866), Macalister (1871), MacDowell (1910), Miller (1952), Ogihara et al. (2005), and according to our dissections. An insertion onto the metacarpal I and trapezium is the common condition in *Pan* according to Gibbs (1999), and in the review of the literature done by Keith (1899) an extension to the proximal phalanx of the thumb was only found in 1 out of 20 *Pan*, while Sarmiento (1994) did not find such an extension in any of the two *Pan* specimens dissected by him,

and in the review of the literature done by Sarmiento (1994) there was no such extension in 10 out of 10 *Pan*.
- Synonymy: Extensor ossis metacarpi plus short extensor of the thumb (Wyman 1855); extensor ossis metacarpi pollicis plus extensor primi internodii pollicis (Wilder 1862, Beddard 1893, Sonntag 1923); extensor ossis metacarpi pollicis (Huxley 1864, Champneys 1872, Sutton 1883, Hepburn 1892, MacDowell 1910, Wood Jones 1920); carpo-métacarpiens du pouce (Gratiolet & Alix 1866); adductor pollicis longus plus extensor pollicis brevis (Ziegler 1964).

CHAPTER 4

Trunk and Back Musculature

Obliquus capitis inferior (LSB 7.8 g; Fig. 23)
- Usual attachments: From the spinous processes of C2 to the transverse process of C1 (Sonntag 1923, Miller 1952).
- Usual innervation: Branch of the dorsal ramus of the first cervical nerve (Miller 1952: *P. paniscus*).

Obliquus capitis superior (LSB 5.5 g; Fig. 23)
- Usual attachments: From the transverse process of C1 to the occipital bone (Sonntag 1923, Miller 1952).
- Usual innervation: Branch of the dorsal ramus of the first cervical nerve (Miller 1952: *P. paniscus*).

Rectus capitis anterior
- Usual attachments: Rectus capitis anterior major runs from C3 to C6 to the basiocciput, receiving a slip from the scalenus anterior, while the rectus capitis anterior minor runs from the anteroventral surface of the lateral mass of C1 to the basiocciput, ventral to the foramen magnum and occipital condyle, and dorsolateral to the insertion of the longus capitis (Sonntag 1923, 1924, Miller 1952, Dean 1985).
- Usual innervation: Data are not available.

Rectus capitis lateralis
- Usual attachments: From the transverse process of C1 to the jugular process of the occipital bone, dorsal to the jugular foramen and lateral to the occipital condyle (Dean 1984).
- Usual innervation: Data are not available.

Rectus capitis posterior major (LSB 8.0 g; Fig. 23)
- Usual attachments: From the spinous process of C2 to the occipital bone, between the inferior nuchal line and the foramen magnum, lateral to the insertion of the rectus capitis posterior minor (Sonntag 1923, Miller 1952).
- Usual innervation: Branches of the dorsal ramus of the first cervical nerve (Miller 1952: *P. paniscus*).

Rectus capitis posterior minor (LSB 6.0 g; Fig. 23)
- Usual attachments: From the dorsal tubercle process of C1 to the medial region of the inferior nuchal line (Sonntag 1923, Miller 1952).
- Usual innervation: Branch of the dorsal ramus of the first cervical nerve (Miller 1952: *P. paniscus*).

Longus capitis (LSB 19.7 g; Fig. 18)
- Usual attachments: From transverse processes of C4 to C7, to the basiocciput (Miller 1952, Dean 1985).
- Usual innervation: Branches of the ventral rami of the upper cervical nerves (Miller 1952: *P. paniscus*).

Longus colli (LSB 26.0 g)
- Usual attachments: The superior oblique division originates from the transverse processes of C3-C6, while the inferior oblique division and the vertical division originate from the body of C6-T3 or C6-T4; the muscle inserts onto C1, C5 and C6 (Sonntag 1923, Miller 1952).
- Usual innervation: Branches of the ventral rami of the cervical nerves (Miller 1952: *P. paniscus*).

Scalenus anterior (LSB 30.2 g; Fig. 18, 20)
- Usual attachments: From the transverse processes of C4, C5 and C6, and occasionally of C7 and/or C3, to the anterior scalene tubercle of the first rib (Champneys 1872, Sonntag 1923, Miller 1952).
- Usual innervation: C8 (Champneys 1872: *P. troglodytes*); lower cervical nerves (Miller 1952: *P. paniscus*).
- Notes: Stewart (1936) stated that chimpanzees and other apes usually have a scalenius anterior and a 'scalenus medius' (which corresponds to the scalenus medius + scalenus posterior *sensu* the present study). According to him the scalenius anterior usually inserts onto the first rib in primates, its origin being limited to C5 and C6 in 'lower primates', but in hominoids the insertion included some of the upper cervical vertebrae. Also according to him, within primates the 'scalenus medius' (i.e., medius + posterior *sensu* the present study) shows a progressive decrease in the number of ribs supplying attachment, usually inserting in hominoids upon the first rib, but occasionally also onto the second rib.

Scalenus medius (LSB 25.4 g; Figs. 18, 20)
- Usual attachments: From C2–C7 to the first and second ribs, the muscle being partly fused with the scalenus posterior (Sonntag 1923, Miller 1952); the combined scalenus medius-scalenus posterior insertion may extend as an aponeurotic sheet to the fifth rib (Sonntag 1923).
- Usual innervation: Lower cervical nerves (Miller 1952: *P. paniscus*).
- Notes: See notes about scalenus anterior above.

Scalenus posterior (LSB 32.3 g; Fig. 18)
- Usual attachments: The combined scalenus medius-scalenus posterior originates from C2–C7 (Miller 1952); the scalenus posterior inserts mainly onto the first (Champneys 1872) and second rib (Macalister 1871), and the combined scalenus medius-scalenus posterior insertion may extend as an aponeurotic sheet to the fifth rib (Sonntag 1923).
- Usual innervation: Lower cervical nerves (Miller 1952: *P. paniscus*).
- Notes: See notes about scalenus anterior above. The **scalenus minimus** is usually not present as a distinct muscle in chimpanzees.

Levatores costarum
- Usual attachments: From the transverse processes of C7–T12 to the costal angle of the next ribs (Miller 1952).
- Usual innervation: Branches of the intercostal nerves (Miller 1952: *P. paniscus*).

Intercostales externi (Figs. 22–23)
- Usual attachments: Attached to the adjacent margins of each pair of ribs (Sonntag 1923, Miller 1952; our specimen VU PT2), being membranous to the sternum in *P. paniscus* (Miller 1952), while in *P. troglodytes* they are replaced by membranous fibrous tissue in all but the first three and last two intercostal spaces (Sonntag 1923).
- Usual innervation: Intercostal nerves (Miller 1952: *P. paniscus*).

Intercostales interni (Figs. 22–23)
- Usual attachments: Connect the adjacent margins of each pair of ribs (our specimen VU PT2).
- Usual innervation: Data are not available.

Transversus thoracis
- Usual attachments: From the internal surface of the xiphoid process and the body of the sternum (Sonntag 1923, Miller 1952, Swindler & Wood 1973), the sternal origin reaches as far anteriorly as the first or second costal spaces (Swindler & Wood 1973), to the inferior border of the costal cartilages of the second to sixth ribs (Sonntag 1923, Miller 1952, Swindler & Wood 1973), and sometimes of the seventh rib (Miller 1952).
- Usual innervation: Intercostal nerves (Miller 1952: *P. paniscus*).

Splenius capitis (LSB 230.7 g; Figs. 18, 20, 23)
- Usual attachments: From the spinous processes of C5–T4 (e.g., Sonntag 1923, Miller 1952) and sometimes of T5–T7 (e.g., Sonntag 1923), to the mastoid process and the occipital bone beneath the superior occipital line (Sonntag 1923, Miller 1952); ocasionally there is a slip to the levator scapulae (Sonntag 1923).
- Usual innervation: Dorsal rami of the cervical nerves (Miller 1952: *P. paniscus*).

Splenius cervicis (LSB 202.0 g; Figs. 18, 20, 23)
- Usual attachments: From T5, receiving a slip from the longissimus cervicis at the level of C2, and inserting onto the transverse process of C1 (Miller 1952).
- Usual innervation: Dorsal rami of the lower cervical nerves (Miller 1952: *P. paniscus*).

Serratus posterior superior (LSB 18.0 g; Fig. 23)
- Usual attachments: From the spinous processes of C7–T1 (Sonntag 1923) and sometimes from C3–T2 (Miller 1952) to ribs 2–5 (Macalister 1871, Sonntag 1923, Miller 1952) and occasionally to rib 1 (Sonntag 1923).
- Usual innervation: Branches of the first four intercostal nerves (Miller 1952: *P. paniscus*).

Serratus posterior inferior (LSB 19.8 g)
- Usual attachments: From the thoracolumbar fascia (Sonntag 1923, Miller 1952) to the inferior four or five ribs, lateral to their angles (Sutton 1883, Sonntag 1923, Miller 1952).
- Usual innervation: Ventral rami of the ninth to twelfth thoracic spinal nerves (Miller 1952: *P. paniscus*).

Iliocostalis (LSB iliocostalis + longissimus + spinalis = 'erector spinae' = 904.3 g; Figs. 20, 23)
- Usual attachments: The costal attachments are onto all the ribs in *P. troglodytes* (Sonntag 1923; our specimen VU PT2), and from all except the first four ribs in *P. paniscus* (Miller 1952), the muscle being fused with the longissimus (Sonntag 1923, Miller 1952). In our *P. troglodytes* specimen VU PT2 the muscle attaches to the iliac crest and some cervical vertebrae, being differentiated into a iliocostalis lomborum, iliocostalis thoracis and iliocostalis cervicis (Fig. 20). The attachments in the cervical region are to the transverse processes of C7 and T1 in *P. paniscus* (Miller 1952).
- Usual innervation: Dorsal rami of the spinal nerves (Miller 1952: *P. paniscus*).
- Notes: The iliocostalis, longissimus and spinalis form the **'erector spinae'** (see Fig. 23).

Longissimus (LSB iliocostalis + longissimus + spinalis = 'erector spinae' = 904.3 g; Figs. 20, 23)
- Usual attachments: In our *P. troglodytes* specimen VU PT2 the muscle mainly runs from iliac crest, sacrum, and adjacent structures, to the ribs (longissimus thoracis), the cervical vertebrae (longissimus cervicis) and the skull (longissimus capitis) (Figs. 20, 23). The cranial insertion of the longissimus is to the occiput accordig to Sonntag (1923) including the mastoid process in *P. paniscus* according to Miller (1952). The cervical insertion is onto the transverse processes of C3 to C5 (Sonntag 1923), extending inferiorly to C6 in *P. paniscus* according to Miller (1952). In the thoracic region, longissimus inserts onto the transverse processes of the thoracic vertebrae according to Sonntag (1923) and Miller (1952) and

inserts onto the ribs between the costal angle and the transverse processes of the thoracic vertebrae, the extent of the costal insertion being from the fourth rib to the last rib according to these authors.
- Usual innervation: Branches of the dorsal rami of the spinal nerves (Miller 1952: *P. paniscus*).
- Notes: The iliocostalis, longissimus and spinalis form the **'erector spinae'** (see Fig. 23). In chimpanzees (e.g., Gratiolet & Alix 1866), modern humans, gorillas and a few other primates the **atlantomastoideus** might be occasionally present as a distinct muscle, running from the atlas to the mastoid process (Diogo & Wood 2012).

Spinalis (LSB iliocostalis + longissimus + spinalis = 'erector spinae' = 904.3 g; Figs. 20, 23)
- Usual attachments: In our *P. troglodytes* specimen VU PT2 the muscle mainly connects the spinous processes of different vertebrae. In specimens dissected by other authors the muscle originates from the spinous processes of T11 and T12 (Sonntag), the origin extending posteriorly to L3 in *P. paniscus* (Miller 1952); the insertion is to the spinous processes of T1 to T8 (Sonntag 1923), also extending anteriorly to C7 in *P. paniscus* (Miller 1952).
- Usual innervation: Branches form the dorsal rami of the spinal nerves (Miller 1952: *P. paniscus*).
- Notes: The iliocostalis, longissimus and spinalis form the **'erector spinae'** (see Fig. 23).

Semispinalis thoracis
- Usual attachments: The origin of the muscle is not separable from the origin of semispinalis cervicis (Sonntag 1923, Winckler 1947; see Semispinalis cervicis below), and was not described in *P. paniscus* by Miller (1952). Its insertion extends may reach as far posteriorly as T8 (Winckler 1947).
- Usual innervation: Data are not available.

Semispinalis cervicis (Fig. 23)
- Usual attachments: From the transverse processes of T1 to T4 (Sonntag 1923, Miller 1952), sometimes extending to T10 and also taking origin from the spinalis muscles (Winckler 1947). The combined origin of semispinalis thoracis and cervicis involves the transverse processes of all the thoracic vertebrae and the articular processes of C2 to C7 according to Sonntag (1923). Winckler (1947) puts the division between thoracis and cervicis at the level of its seventh fascicle, which arises from T7 to T10 and inserts onto T1. The insertion of the semispinalis cervicis is onto the spinous processes of C1 to C6 (Sonntag 1923, Miller 1952), reaching posteriorly as far as T1 according to Winckler (1947).
- Usual innervation: Dorsal rami of the cervical nerves (Miller 1952: *P. paniscus*).

Semispinalis capitis
- Usual attachments: Origin is from the articular processes of C4 to T4 (Miller 1952), often extending anteriorly to C3 (Sonntag 1923, Winckler 1947) and posteriorly to T6 (Sonntag 1923) or T7 (Winckler 1947), and often fused with the longissimus thoracis and the spinalis thoracis at its origin (Sonntag 1923). Insertion of the semispinalis capitis is onto the occipital bone between the superior and inferior nuchal lines (Sonntag 1923, Miller 1952).
- Usual innervation: Dorsal rami of the cervical nerves (Miller 1952: *P. paniscus*).

Multifidus
- Usual attachments: Origin is from the sacrum and sacroiliac ligaments, the mammillary processes of the lumbar certebrae, the transverse processes of the thoracic vertebrae, and the articular processes of C7 to C4 (Sonntag 1923, Winckler 1947), sometimes extending anteriorly to C3 (Miller 1952) or to C2 (Winckler 1947). Insertion is onto the lamina and the spinous processes of the lumbar and thoracic vertebrae and to the cervical vertebrae as far as, and including, C2 (Sonntag 1923, Winckler 1947) or C1 (Miller 1952).
- Usual innervation: Dorsal rami of the spinal nerves (Miller 1952: *P. paniscus*).

Rotatores
- Usual attachments: From the transverse processes of the thoracic vertebrae to the lamina and base of the spinous process of the vertebra above ('rotatores breves', the 'rotatores longi' being seemingly absent: Sonntag 1923).
- Usual innervation: Data are not available.
- Synonymy: Rotatores breves and longi (Gibbs 1999).

Interspinales
- Usual attachments: To our knowledge, there are no detailed descriptions of these muscles in chimpanzees, the only mention to them being Sonntag (1923), who stated that they are similar to those of modern humans.
- Usual innervation: Data are not available.

Intertransversarii
- Usual attachments: To our knowledge, there are no detailed descriptions of these muscles in chimpanzees, the only mention to them being Sonntag (1923), who stated that they are similar to those of modern humans.
- Usual innervation: Data are not available.

CHAPTER 5

Diaphragmatic and Abdominal Musculature

Diaphragma (89.9 g unpaired muscle, so each side = 44.95 g; Fig. 22)
- Usual attachments: See Notes below.
- Usual innervation: Phrenic nerve (Miller 1952: *P. paniscus*).
- Notes: The position of the aortic aperture is higher than in lower catarrhines and is shorter by two-thirds of a vertebra in modern humans than in *Pan* (Juraniec 1972). The hiatus extends posteriorly on average to the level of the upper part of L3, and anteriorly to T11 or T12 (Juraniec 1972). The division of the central tendon into folia is as distinct in *Pan* as it is in modern humans. In 1/3 *Pan* the central tendon is not divided into leaflets, being kidney-shaped (Juraniec & Szostakiewicz-Sawicka 1968). In 2/3 *Pan* the overall shape of the central tendon is similar to that of modern humans, with lateral and anterior leaflets of approximately the same size (Juraniec & Szostakiewicz-Sawicka 1968). The central point of decussation is present (Blair 1923), and muscle fibres are found in the anterior leaflet in 1/3 *Pan* (Juraniec & Szostakiewicz-Sawicka 1968). The oesophageal hiatus is situated in the muscular part of the diaphragm at the level of T11, and the middle of T10 (Juraniec 1972). It is elliptical in shape (Juraniec 1972), and is formed by the splitting of the medial fibres of the right crus (Juraniec 1972, Körner 1929, Low 1907, Miller 1952). The oesophageal and aortic hiatuses are separated by half a vertebra (Juraniec 1972). Juraniec (1972) distinguished four types of fibres crossing at the oesophageal hiatus in primates (in type I, only muscular crossing is present, with 6 subtypes distinguished). Type I4, I5 and I6 represent various types of crossing of the fibres of the crura to the opposite side in bundles. These account for 1/4 *Pan*. In type III, not found in *Homo*, a tendinous strand at the site of the crossing of the fibres is present on the right side of a *P. troglodytes* specimen dissected by Juraniec (1972) (subtype III2). The pars costalis of the diaphragm originates from the lower seven ribs and interdigitates with the transverses abdominis (Sonntag 1923, Miller 1952). The pars lumbaris originates from the

lumbar vertebrae by two crura (Sonntag 1923, Miller 1952); additional slips arise from the transverse process of L2 and from the side of the body or L1 (Sonntag 1923). This pars lumbaris also originates from two medial and lateral arcuate ligaments and is divided into a right crus arising from the bodies and intervertebral discs as far posteriorly as L2 (Sonntag 1923, Miller 1952) and a left crus arising from L1 and L2 in *P. paniscus* (Miller 1952) and from L1 in *P. troglodytes* (Sonntag 1923). The *lateral arcuate ligament* is a thickened band in the quadrates lumborum fascia that attaches to the transverse process of L1 (Miller 1952). It also attaches to the inferior margin to the last two ribs (ribs thirteen and fourteen) Miller 1952. The medial arcuate ligament is a tendinous arch in the psoas major fascia and attaches to the side of the body of L1, being laterally fixed to the front of the transverse process of L1 (Miller 1952). The pars sternalis originates by two slips from the inner surface of the xiphoid process in *P. paniscus* (Miller 1952), while in *P. troglodytes* it takes origin by two slips from the inner surface of the sternum (Sonntag 1923).

Rectus abdominis (LSB 298.8 g; Figs. 19, 21–22)
- Usual attachments: From the outer surface of the costal cartilages of the fifth to seventh ribs to the region of the pubic crest (Miller 1952); there are four tendinous intersections (Champneys 1872, Sonntag 1923, Miller 1952), of which two are posterior to the umbilicus (Miller 1952). The superficial rectus sheath is formed by the external and internal oblique, and the deep sheath by the transversus abdominis alone (Champneys 1872, Lunn 1949, Miller 1952).
- Usual innervation: Lower intercostals nerves in *Pan paniscus* Miller 1952.
- Notes: To our knowledge, the presence of a **supracostalis** and **tensor linea semilunaris** has not been reported in chimpanzees.

Pyramidalis (LSB 13.7 g; Fig. 21)
- Usual attachments: In her review of the literature, Gibbs (1999) suggested that the pyramidalis is probably usually missing in chimpanzees, being absent in the *P. troglodytes* specimens reported by Champneys (1872), Sonntag (1923), Vrolik (1841). It was present on the left side of the *P. paniscus* specimen reported by Miller (1952)—it ran from the pubic crest between the rectus abdominis and the rectus sheath, radiating into the linea alba—but it was absent on the right side. In the single *P. troglodytes* in which we analysed this feature in detail (VU PT1) the muscle was present as a small and triangular structure that lies superficial (ventral) to the rectus abdominis running from the pelvic region to the region of the linea alba (Fig. 21).
- Usual innervation: Twelfth thoracic nerve (Miller 1952: *P. paniscus*).
- Notes: This muscle is present in about 20% of modern humans, which in general display a configuration similar to that found in our *P. troglodytes* specimen.

Cremaster
- Usual attachments: From the internal abdominal oblique to the 'inguinal ligament' (see Notes below), also receiving a contribution from the transversus abdominis (Mijsberg 1923, Miller 1947, 1952, Lunn 1949).
- Usual innervation: Data are not available.
- Notes: According to the literature review of Gibbs (1999) there is no true inguinal ligament in any ape, only a series of tendinous arches over the sartorius and the femoral vessels and nerves, merging with the fascia lata (Glidden & De Garis 1936, Miller 1947, 1952, Lunn 1948). Schultz (1936), Miller (1947) and Lunn (1948) suggest that a true inguinal ligament is only found in modern humans, but ligament-like connective tissue in a plane ventral to the pelvic girdle has been identified in all of the mammals dissected by Lunn (1948).

Obliquus externus abdominis (LSB 234.0 g; Figs. 19, 21, 25)
- Usual attachments: Originates by fleshy slips from the external surface of the sixth to eleventh ribs (Champneys 1872, Sonntag 1923, Winckler 1950), extending posteriorly to the last rib in *P. paniscus* (Miller 1952). The anterior extent of the origin is also variable in *Pan* (Champneys 1872, Sonntag 1923, Winckler 1950). The anterior part of the muscle interdigitates with slips of the serratus anterior and the posterior part with slips of the latissimus dorsi (Champneys 1872, Sonntag 1923), while its medial part interdigitates with the pectoralis major (Sonntag 1923). Insertion of the fibres from the last ribs is onto the iliac crest and its lateral lip, as far as the anterior superior iliac spine (Champneys 1872, Kohlbrügge 1897, Sonntag 1923, Waterman 1929, Miller 1952). From the anterior superior iliac spine to the pubic tubercle, the muscle has a free posterior aponeurotic border (Kohlbrügge 1897, Sonntag 1923, Lunn 1949, Miller 1952).
- Usual innervation: Branches of the lower intercostal nerves and also by the iliohypogastric and ilioinguinal nerves (Miller 1952: *P. paniscus*).

Obliquus internus abdominis (LSB 191.7 g)
- Usual attachments: Origin is from the anterior part of the iliac crest and from the deep layer of the thoracolumbar fascia (Miller 1952), the aponeurosis of the external oblique, or the 'inguinal ligament' (Sonntag 1923, Miller 1947, 1952, Lunn 1949). The cranial part of the muscle inserts onto the inferior margins of the cartilage of the last four ribs (Sonntag 1923, Miller 1952); the middle fibres become aponeurotic medially, forming the anterior wall of the rectus sheath (Champneys 1872, Lunn 1949, Miller 1952).
- Usual innervation: Branches of the lower intercostal nerves and the first lumbar nerve (Miller 1952: *P. paniscus*).

Transversus abdominis (LSB 180.0 g; Fig. 22)
- Usual attachments: This muscle forms the posterior layer of the rectus sheath (Champneys 1872, Lunn 1949, Miller 1952). Its fibres decussate in the linea

alba, and the aponeurotic part forms, along with the internal oblique, the conjoined tendon, which inserts into the superior pubic surface in the region of the pubic crest (Sonntag 1923, Miller 1947, 1952, Lunn 1949).
- Usual innervation: Lower intercostal nerves and first lumbar nerve (Miller 1952: *P. paniscus*).

Quadratus lumborum (LSB 65.5 g; Fig. 22)
- Usual attachments: From the medial lip of the iliac crest and the iliolumbar ligament, its ventral fibres being continuous with the iliacus, to the penultimate rib and also to the transverse processes of L1–L3 or L1–L4 (Miller 1952).
- Usual innervation: Ventral rami of all of the lumbar nerves (Miller 1952: *P. paniscus*).

CHAPTER 6

Perineal, Coccygeal and Anal Musculature

Coccygeus
- Usual attachments: The muscle is mostly tendinous with few muscle fibres inserting onto the side of the coccyx, its posterior border being continous with the pubococcygeus (Elftman 1932, Miller 1952), or its insertion with the coccygeal part of the origin of the gluteus maximus (Champneys 1872). According to Elftman (1932) in chimpanzees there is no trace of the **flexor caudae** and according to Elftman (1932) and Sonntag (1923) the **iliococcygeus** is also usually absent.
- Usual innervation: Third sacral nerve (Champneys 1872: *P. troglodytes*); fourth and fifth sacral nerves (Miller 1952: *P. paniscus*).

Levator ani
- Usual attachments: The muscle is formed by bilateral 'plates' of muscle and fibrous tissue that have a linear attachment to the obturator fascia on the inner wall of the lesser pelvis that extends from the pubic symphysis to the ischial ramus (Thompson 1901, Sonntag 1923, Miller 1952). The posteriorly directed fibres converge in the midline; the more anterior fibres encircle the rectum and the more posterior ones insert onto the anococcygeal raphe behind the anus as well as onto the tip of the coccyx (Sonntag 1923, Miller 1952).
- Usual innervation: Branches of the rectal nerve from the fourth and fifth sacral nerves (Miller 1952: *P. paniscus*).
- Notes: To our knowledge, the ***pubovesicalis*** ('ligamentum puboprostaticum' or 'puboampullaris': Gibbs 1999) has never been reported in chimpanzees.

Pubococcygeus
- Usual attachments: Origin is from the posterior surface of the pubic body (Thompson 1901, Elftman 1932), extending to the ischial spine (Smith 1923, Elftman 1932) and also originating from the obturator fascia (Thompson 1901, Van den Broek 1908). Insertion is onto the rectal wall (Thompson 1901; variably in *Pan* according to Elftman 1932) as well as onto the tip of the coccyx

(Smith 1923, Elftman 1932), the anococcygeal raphe (Elftman 1932), and/or the sacrum (Smith 1923). The pubococcygeus muscles of the two sides unite at the midline (Thompson 1901, Smith 1923, Elftman 1932).
- Usual innervation: Data are not available.
- Notes: According to Elftman (1932) and Thompson (1901) in great apes the **puborectalis** is homologous with the inferior fibres of the pubococcygeus.

Sphincter ani externus
- Usual attachments: The sphincter encircles the anus, extending from the anococcygeal raphe (Sonntag 1923, Miller 1952) to the perineal body (Elftman 1932, Miller 1952) and forming a muscular basin (less marked degree in *Pan* than in *Gorilla* and *Pongo*) for the support of both alimentary and urogenital viscera according to Elftman 1932. Some deeper fibres blend with the inferior margin of the levator ani (Sonntag 1923, Elftman 1932).
- Usual innervation: Branches of the fourth and fifth sacral nerves (Miller 1952: *P. paniscus*).

Bulbospongiosus
- Usual attachments: Origin is from the median raphe of the penile bulb and from the ischium, the muscle surrounding the bulb and the corpora of the penis (Elftman 1932).
- Usual innervation: Data are not available.
- Notes: Some fibres of the bulbospongiosus that lie in contact with the pelvic diaphragm and insert into the ischium in chimpanzees may be homologous with the **transversus perinei profundus** according to Elftman (1932). According to Sonntag (1923) the *transversus perinei superficialis* is absent in *Pan*.
- Synonymy: Bulbocavernosus (Gibbs 1999).

Ischiocavernosus
- Usual attachments: Origin is from the ascending pubic ramus (Sonntag 1923); no information about insertion.
- Usual innervation: Data are not available.

Sphincter urethrae
- Usual attachments: In female *Pan*, its fibres interlace in the vaginal wall (Roberts & Seibold 1971).
- Usual innervation: Data are not available.

Rectococcygeus
- Usual attachments: To our knowledge there are no detailed descriptions of this muscle in chimpanzees.
- Usual innervation: Data are not available.
- Synonymy: Caudoanalis (Gibbs 1999).

Rectourethralis

- Usual attachments: To our knowledge there are no detailed descriptions of this muscle in chimpanzees.
- Usual innervation: Data are not available.
- Notes: To our knowledge the muscles **rectovesicalis, regionis analis, regionis urogenitalis, compressor urethrae, sphincter urethrovaginalis, sphincter pyloricus, suspensori duodeni, sphincter ani internus, sphincter ductus choledochi, sphincter ampullae, detrusor vesicae, trigoni vesicae, vesicoprostaticus, vesicovaginalis, puboprostaticus** and **rectouterinus** have not been described in detail in chimpanzees. We were unable to examine these muscles in detail in the chimpanzees dissected by us.

CHAPTER 7

Pelvic and Lower Limb Musculature

Iliacus (Fig. 22)
- Usual attachments: From the iliac fossa (Hepburn 1892) or the entire ventral surface of the ilium (Sigmon 1974) to the medial aspect of the lesser trochanter of the femur, in combination with the psoas major (Champneys 1872, Beddard 1893).
- Usual innervation: Branches of femoral nerve (Sigmon 1974: *P. troglodytes*).

Psoas major (LSB 179.1 g; Fig. 22)
- Usual attachments: From the lateral surfaces of the bodies and the costal processes of the lumbar vertebrae (Champneys 1872, Hepburn 1892), extending proximally to T12 in some specimens (Champneys 1872, Sigmon 1974) and distally to S1 in most *Pan* (Hepburn 1892, Sigmon 1974). It may also take origin from the intervening intervertebral dics (Hepburn 1892, Sigmon 1974) and/or from the head of the twelfth rib (Champneys 1872).
- Usual innervation: Femoral nerve and first two or three lumbar nerves (Champneys 1872, Hepburn 1892, Sonntag 1923, Sigmon 1974: *P. troglodytes*).

Psoas minor (LSB 15.3 g; Fig. 22)
- Usual attachments: Originates mainly from the ventrolateral surface of L1 (Champneys 1872, Hepburn 1892, Swindler & Wood 1973). An origin from the last thoracic vertebra is present in half of the *Pan* specimens dissected by Champneys (1872), Hepburn (1892), Sigmon (1974) and Swindler & Wood (1973). An origin from L2 (Hepburn 1892) and/or from the intervertebral discs (Swindler & Wood 1973) may also be present. Insertion is onto the iliopubic eminence and the pectineal line (Champneys 1872, Hepburn 1892, Swindler & Wood 1973).
- Usual innervation: First lumbar nerve (Champneys 1872, Hepburn 1892: *P. troglodytes*), but it may also be innervated by the twelfth thoracic nerve Hepburn (1892) and Sigmon (1974) (*P. troglodytes*).

Gluteus maximus (LSB 187.4 g; RSB 212.7 g; Figs. 20, 41)
- Usual innervation: From the posterior iliac crest, thoracolumbar fascia, sacrum, coccyx, the sacrotuberous ligament and the fascia over gluteus medius

(Hepburn 1892, Beddard 1893, Ranke 1897, MacDowell 1910, Van den Broken 1914, Mysberg 1917, Miller 1945, Robinson et al 1972, Sigmon 1974 Brown 1983, Hamada 1985). There is often also an origin from the ischial tuberosity (Champneys 1872, Hepburn 1892, Beddard 1893, Dwight 1895, MacDowell 1910, Robinson et al. 1972, Sigmon 1974, Zihlman & Brunker 1979, Brown 1983, Hamada 1985). The muscle may share its origin with the long head of the biceps femoris (Champneys 1872), and may be subdivided into as many as three (Hamada 1985) or four (our specimen GWUANT PT1) parts. The gluteus maximus inserts onto the iliotibial tract (when present) and the posterolateral aspect of the femur in the region of the gluteal tuberosity, and occasionally onto the hypertrochanteric fossa on the lateral aspect of the femur (Humphry 1867, Champneys 1872, MacDowell 1910, Sigmon 1974, Brown 1983, Hamada 1985). The insertion of the muscle is generally extends more distally on the femur in apes than in modern humans, extending almost to the lateral condyle of the femur (Dwight 1895, Hepburn 1892, MacDowell 1910, Sigmon 1974, Zihlman & Brunker 1979, Hamada 1985).

- Usual innervation: Data are not available.

Gluteus medius (LSB 538.9 g; RSB 438.2 g; Figs. 20, 41)
- Usual attachments: From the lateral surface of the ilium and the gluteal fascia (Champneys 1872) and also from the fascia lata (Beddard 1893), to the lateral aspect of the greater trochanter (Champneys 1872), its tendon of insertion being often split by the vastus lateralis (Champneys 1872, Hepburn 1892, Sonntag 1924) and fused with that of the piriformis (Champneys 1872, Hepburn 1892, Dwight 1895).
- Usual innervation: Superior gluteal nerve (Sigmon 1974: *P. troglodytes*).

Gluteus minimus (LSB 21.9 g; RSB 66.9 g; Figs. 41–42)
- Usual attachments: From the dorsolateral surface of the ilium, extending from just distal to the anterior superior iliac spine, towards the acetabulum (Champneys 1872, Sigmon 1974) and sometimes from the margin of the greater sciatic notch (Beddard 1893) and/or the ischial spine (Sigmon 1974), to the greater trochanter (Champneys 1872, Hepburn 1892, Sigmon 1974). The origin is in two parts according to Champneys (1872). The muscle is triangular or fan-shaped (Sigmon 1974) and it is usually larger in African apes than in the Asian apes (Hepburn 1892).
- Usual innervation: Superior gluteal nerve (Sigmon 1974: *P. troglodytes*).

Scansorius
- Usual attachments: The scansorius is occasionally present in *Pan* (Wilder 1862, Champneys 1872, Hepburn 1892, Beddard 1893, Blake 1976), running from the anterolateral ilium adjacent to the acetabular rim (Beddard 1893, Dwight 1895, Sonntag 1923) and sometimes from the fascia lata (Beddard 1893), to the greater trochanter, distal to the insertion of the gluteus minimus (Wilder

1862, Champneys 1872, Hepburn 1892, Beddard 1893, Sonntag 1923). When present, it is a flat triangular muscle (Wilder 1862, Beddard 1893, Sigmon 1974) that is often blended with the gluteus minimus (Champneys 1871, Hepburn 1892, Dwight 1895).
- Usual innervation: Superior gluteal nerve (Champneys 1872: *P. troglodytes*).
- Notes: We examined this region in detail in three chimpanzees (specimenss GWUANT PT1, GWUANT PT2 and VU PT1); in each the scansorius was missing.

Ischiofemoralis (LSB 313.9 g; RSB 303.3 g; Fig. 41, 45, 52)
- Usual attachments: Although initially described by some authors as comprising the distal third of the gluteus maximus (Stern 1972, Sigmon 1974, Hamada 1985), the ischiofemoralis is now considered to be an independent muscle, particularly in light of the innervation studies of Kaseda et al. (2008). This muscle is usually missing in modern humans but it is particularly developed in *Pongo*. In apes the muscle runs from the ischial tuberosity to the lateral aspect of the femoral shaft and the aponeurosis of the vastus lateralis. In the chimpanzees dissected by us the ischiofemoralis is blended with the gluteus maximus, running from the ischial tuberosity to the femoral diaphysis and the aponeurosis of the vastus lateralis.
- Usual innervation: Common fibular component of the sciatic nerve (Kaseda et al. 2008: *P. troglodytes*).

Gemellus superior (LSB 4.4 g; RSB 2.0 g; Fig. 42)
- Usual attachments: Runs from the region of the ischial spine between the ischial spine and the ischial tuberosity (Sigmon 1974) to the trochanteric fossa together with the tendon of the obturator internus, with which it is often fused (Champneys 1872, Beddard 1893, Sigmon 1974).
- Usual innervation: Branches of sacral plexus (Champneys 1872, Sigmon 1974: *P. troglodytes*).

Gemellus inferior (LSB 5.9 g; RSB 7.9 g; Fig. 42)
- Usual attachments: Originates from the region of the ischial tuberosity (Champneys 1872, Hepburn 1892) to the trochanteric fossa of the femur (Champneys 1872, Beddard 1893, Sigmon 1974), being difficult to isolate from the quadratus femoris (Hepburn 1892) and/or from the obturator internus (our dissections).
- Usual innervation: Branches of sacral plexus (Champneys 1872, Sigmon 1974: *P. troglodytes*).

Obturatorius externus (LSB 71.7 g; RSB 54.4 g)
- Usual attachments: It runs from the external surface of the medial bony margin of the obturator foramen and from the obturator membrane (Sigmon 1974) and in our specimens also from the femoral neck, fusing with the obturator internus (Hepburn 1892) before it inserts with it onto the trochanteric fossa

(Hepburn 1892, Sigmon 1974); occasionally it also inserts onto the lesser femoral trochanter (Beddard 1893).
- Usual innervation: Obturator nerve (Sigmon 1974: *P. troglodytes*).
- Synonymy: Obturator externus (Gibbs 1999).

Obturatorius internus (LSB 72.0 g; RSB 38.7 g; Fig. 42)
- Usual attachments: It originates from the margin of the obturator foramen and the obturator membrane, extending to the descending pubic ramus and the medial surface of the inferior ischial ramus (Champneys 1872, Sigmon 1974, Swindler & Wood 1973). The origin may extend to the ischial tuberosity and pubic symphysis, and a tendinous arch running across the obturator foramen (Champneys 1872). The muscle runs through the lesser sciatic foramen (Hepburn 1892) and inserts together with the gemelli muscles onto the trochanteric fossa of the fermur (Champneys 1872, Beddard 1893, Sigmon 1974).
- Usual innervation: Sacral nerves (Sigmon 1974: *P. troglodytes*).
- Synonymy: Obturator internus (Gibbs 1999).

Piriformis (LSB 30.0 g; RSB 44.5 g; Fig. 41)
- Usual attachments: It originates by slips (two well-differentiated heads of origin in our specimen GWUANT PT1) from the anterolateral surface of the distal half of the sacrum (Champneys 1872, Hepburn 1892, Sigmon 1974), the posterior extent of the origin often being S5 (Champneys 1872, Hepburn 1892). An origin from the greater sciatic notch is often present (Chapman 1879, Sigmon 1974), and the muscle is often fused with the gluteus medius (Champneys 1872, Hepburn 1892, Dwight 1895, Sigmon 1974). It passes through the greater sciatic foramen to insert onto the ventromedial aspect of the tip of the greater trochanter of the femur (Champneys 1872, Hepburn 1892, Beddard 1893, Sigmon 1974).
- Usual innervation: Branches of sacral plexus (Sigmon 1974: *P. troglodytes*).
- Notes: In the three chimpanzees in which we examined this region in detail (GWUANT PT1, GWUANT PT2 and VU PT1) the piriformis is not fused with the gluteus medius.

Quadratus femoris (LSB 11.0 g; RSB 14.1 g; Fig. 42)
- Usual attachments: It originates from the ventrolateral aspect of the ischial tuberosity (Beddard 1893, Dwight 1895, Miller 1945, Sigmon 1974), being a small, thick, fleshy muscle that is separated from the adductor magnus and closely related to the inferior gemellus (Hepburn 1892). It mainly inserts onto the intertrochanteric crest of the femur (Hepburn 1892, Dwight 1895, Sigmon 1974), sometimes being also attached to the dorsal surface of the greater trochanter (Hepburn 1892) and/or dorsal to the lesser trochanter (Champneys 1872, Dwight 1895, Hepburn 1892).
- Usual innervation: Sacral nerves (Sigmon 1974: *P. troglodytes*).

- Notes: Hepburn (1892) stated that the **articularis genu** is present in all apes, but the muscle is usually not reported in chimpanzees (e.g., Sonntag 1923, 1924) and in the chimpanzees in which we examined this anatomical region in detail (GWUANT PT1, GWUANT 2 and VU PT1) the articularis genu is seemingly not present as a distinct muscle.

Rectus femoris (LSB 140.7 g; RSB 132.8 g; Figs. 42–43)
- Usual attachments: It possesses two heads of origin in 1/3 of the chimpanzees dissected by Dwight (1895), Hepburn (1892) and Vrolik (1841). The short head, usually present, originates from the anterior inferior iliac spine (Champneys 1872). The reflected head, when present, originates from the ilium near the acetabulum (Hepburn 1892, Beddard 1893, Boyer 1935, Sigmon 1974), an origin that is similar to that of the single head of origin found in our specimens VU PT1 and GWUANT PT1. In a single specimen of Pan, rectus femoris was fused with vastus lateralis (Beddard 1893). It inserts together with the other muscles forming the 'quadriceps femoris' (i.e., vastus intermedius, vastus laterais and vastus medialis), mainly to the proximolateral border of the patella, and then via the patellar tendon to the tibial tuberosity (Beddard 1893, Hepburn 1892).
- Usual innervation: Data are not available.

Vastus intermedius (LSB 205.3 g; RSB 240.8 g; Fig. 42)
- Usual attachments: Originates from the anterior femoral shaft (Sigmon 1974). It inserts together with the other muscles forming the 'quadriceps femoris' (i.e., rectus femoris, vastus laterais and vastus medialis), mainly to the proximolateral border of the patella, and then via the patellar tendon to the tibial tuberosity (Beddard 1893, Hepburn 1892).
- Usual innervation: Data are not available.

Vastus lateralis (LSB 592.4 g; RSB 437.4 g; Figs. 42, 45)
- Usual attachments: It originates from the lateral aspect of the greater femoral trocanter and the distal two-thirds of the lateral femoral shaft, in the region of the lateral lip of the linea aspera (Beddard 1893, Sigmon 1974), and occasionally from the iliofemoral ligament (Sigmon 1974). It inserts together with the other muscles forming the 'quadriceps femoris' (i.e., rectus femoris, vastus intermedius and vastus medialis) mainly to the proximolateral border of the patella, and then via the patellar tendon to the tibial tuberosity (Beddard 1893, Hepburn 1892).
- Usual innervation: Data are not available.

Vastus medialis (LSB 293.8 g; RSB 236.4 g; Figs. 42–44)
- Usual attachments: It originates from the dorsomedial femoral shaft, in the region of the linea aspera (Champneys 1872, Beddard 1893, Sigmon 1974) and often also from the iliofemoral ligament (Beddard 1893, Sigmon 1974) and/or from the femoral neck (our specimens VU PT1 and GWUANT PT1). It inserts

together with the other muscles forming the 'quadriceps femoris' (i.e., rectus femoris, vastus laterais and vastus intermedius) mainly to the proximolateral border of the patella, and then via the patellar tendon to the tibial tuberosity (Beddard 1893, Hepburn 1892).
- Usual innervation: Data are not available.

Sartorius (LSB 115.9 g; RSB 122.4 g; Figs. 43, 45)
- Usual attachments: It originates from the anterior iliac border (Humphry 1867, Champneys 1872, Hepburn 1892, Beddard 1893), often stated as the region of the anterior superior iliac spine (Sigmon 1974). The ribbon-shaped muscle courses obliquely over the thigh dorsal to the medial femoral condyle to insert onto the medial border of the tibial shaft, its insertion being often superficial to those of the gracilis and semitendinosus (Chapman 1979, Hepburn 1892, Beddard 1893, Sigmon 1974).
- Usual innervation: Femoral nerve (Sigmon 1974: *P. troglodytes*).

Tensor fasciae latae (LSB 86.1 g whole muscle; RSB 12.0 g muscle belly only; Fig. 45)
- Usual attachments: Runs from the region of the anterior superior iliac spine (Humphry 1867, Champneys 1872, Kaplan 1958a, Sigmon 1974) and often from the gluteal fascia (Kaplan 1958a, Sigmon 1974), to the iliotibial tract (Duvernoy 1855-1856, Hepburn 1892, Kaplan 1958a, Sigmon 1974).
- Usual innervation: Superior gluteal nerve (Sigmon 1974: *P. troglodytes*); inferior gluteal nerve (Miller 1952: *P. paniscus*).

Adductor brevis (LSB 135.0 g; RSB 91.4 g; Fig. 44)
- Usual attachments: Originates from the body of the pubis (Hepburn 1892, Beddard 1893, Dwight 1895, Sigmon 1974), sometimes also from the inferior pubic ramus near the symphysis (Beddard 1893) and/or from the superior pubic ramus (Sigmon 1974), and occasionally from the ischium and intermuscular septa (Beddard 1893). The muscle may be divided into two parts (Champneys 1872, Beddard 1893, Hepburn 1892) and is often partly fused with the short head of the adductor magnus (Hepburn 1892, Sigmon 1974). In a single *Pan* specimen the adductor muscles were described as being indistringuishable from each other, forming an "adductor mass" (Dwight 1895). The insertion of the adductor brevis is distal to the lesser trochanter onto the proximal third of the medial lip of the linea aspera on the mid-dorsal femoral surface, onto the pectineal line (Champneys 1872, Hepburn 1892, Beddard 1893, Dwight 1895, Sigmon 1974).
- Usual innervation: Ventral ('anterior') division of the obturator nerve (Champneys 1872, Sigmon 1974, Hamada 1985: *P. troglodytes*).

Adductor longus (LSB present but undetermined; RSB 389.1 g; Fig. 43)
- Usual attachments: Originates by a flat tendon from the anterior superior pubic ramus in the region of the pubic tubercle (Humphry 1867, Beddard 1893, Dwight 1895, Sigmon 1974) and occasionally from the intermuscular septum

(Beddard 1893). It inserts onto the middle of the medial lip of the linea aspera or mediodorsal femoral shaft (Humphry 1867, Hepburn 1892, Beddard 1893, Dwight 1895, Sigmon, 1974), besides and ventral to the proximal half of the insertion of the short head of the adductor magnus (Beddard 1893, Sigmon 1974). There may be an expansion to the medial condyle (Humphry 1867), and the insertion of the adductor longus may be fused with the adductor magnus (Champneys 1872, Beddard 1893).
- Usual innervation: Data are not available.

Adductor magnus (LSB 410.6 g; RSB 585.8 g; Fig. 44)
- Usual attachments: From the ventral surface of the inferior pubic ramus, lateral to the symphysis, and from the inferior ischial ramus as far as the ischial tuberosity (Humphry 1867, Champneys 1872, Hepburn 1892, Beddard 1893, Dwight 1895, Sigmon 1974, Yirga 1987), and occasionally from the inferiomedial border of the semitendinosus and the long head of the biceps femoris (Sigmon 1974). Its insertion is onto the medial lip of the linea aspera or the dorsomedial surface of the femur, and to the adductor tubercle of the medial epicondyle (Humphry 1867, Champneys 1872, Hepburn 1892, Beddard 1893, Dwight 1895, Sigmon 1974). Its short head often also attaches onto the lateral border of the common tendon of insertion of the adductor longus and adductor brevis (Champneys 1872, Beddard 1893).
- Usual innervation: Short head by the obturator nerve (Champneys 1872, Hepburn 1892, Sigmon 1974: *P. troglodytes*); long head by flexores femoris nerve (Sigmon 1974: *P. troglodytes*) or by sciatic nerve (Champneys 1872, Hepburn 1892: *P. troglodytes*); according to Sigmon (1974) the flexor femoris nerve does not exist in modern humans, and the tibial nerve in apes (except in a few gorillas) does not supply the adductor magnus.
- Notes: According to Sigmon (1974) the flexor femoris nerve does not exist in modern humans, and the tibial nerve in apes (except in a few gorillas) does not supply the adductor magnus.

Adductor minimus
- Usual attachments: The adductor minimus is absent from a third of great apes (Hepburn 1892, Sigmon 1974), being often described as a superior subdivision of the adductor magnus or as a deep slip of the adductor brevis (Hepburn 1892). When present it runs from the inferior pubic ramus (Hepburn 1892, Dwight 1895) to the linea aspera on the dorsal femoral shaft (Hepburn 1892).
- Usual innervation: Data are not available.

Gracilis (LSB 248.3 g; RSB 253.2 g; Figs. 43, 52)
- Usual attachments: From the inferior pubic ramus near to the pubic symphysis onto the ischial ramus (Champneys 1872, Hepburn 1892, Beddard 1893, Sigmon 1974) and sometimes from the whole pubic body (Hepburn 1892) or from the superior pubic ramus (Sonntag 1924). Insertion is to the ventromedial

surface of the tibia (Humphry 1867, Champneys 1872, Hepburn 1892, Dwight 1895, Sigmon, 1974), and often through an aponeurotic expansion to the fascia of the leg (Champneys 1872, Hepburn 1892, Dwight 1895).
- Usual innervation: Ventral ('anterior') branch of obturator nerve (Sigmon 1974: *P. troglodytes*).

Pectineus (LSB 90.8 g; RSB 39.9 g; Figs. 43–44)
- Usual attachments: From the superior pubic ramus (Hepburn 1892, Sigmon 1974) to the dorsal femur just distal to the lesser trochanter (Hepburn 1892, Beddard 1893, Sigmon 1974).
- Usual innervation: Femoral nerve and occasionally also by the ventral 'anterior' branch of the obturator nerve (Champneys 1872, Sigmon 1974: *P. troglodytes*).
- Notes: To our knowledge, the **iliocapsularis** was not described as a distinct muscle in chimpanzees.

Biceps femoris (LSB caput longum 163.0 g; RSB caput longum 157.9 g; LSB caput breve 125.8 g; RSB 136.3 g; Figs. 41, 45, 52)
- Usual attachments: The long head of the biceps femoris originates from the ischial tuberosity, often in common with the semitendinosus (Champneys 1872, Beddard 1893, Prejzner-Morawska & Urbanowick 1971, Sigmon 1974, Hamada 1985, Kumakura 1989) and less often with the semimembranosus (Champneys 1872, Beddard 1893), the gluteus maximus (Champneys 1872, Sigmon 1974), and/or the posterior part of the gracilis (Champneys 1872). The short head originates from the dorsdolateral femur in the region of the lateral lip of the linea aspera (Beddard 1893, Prejzner-Morawska & Urbanowick 1971, Sigmon 1974, Hamada 1985, Kumakura 1989), extending more distally in great apes than in *Hylobates* and modern humans (e.g., Sigmon 1974). The short head is absent from a single specimen of *Pan* (Sutton 1883), the two heads being fused in 21/26 chimpanzees according to the literature review of Gibbs (1999). In our specimen VU PT1 there is an additional belly of the short head of the biceps femoris inserting onto the tendon of insertion of the long head of this muscle (Fig. 52). The insertion of the long head of biceps femoris is usually to the tibial head (Hepburn 1892, Prejzner-Morawska & Urbanowick 1971, Sigmon 1974), and to the tibial tuberosity, or condyle (Champneys 1872, Dwight 1895, Sonntag 1923, Kaplan 1958a,b, Prejzner-Morawska & Urbanowick 1971, Sigmon 1974) and often to the fibular head and fascia of the leg (Champneys 1872, Beddard 1893, Dwight 1895, Prejzner-Morawska & Urbanowick 1971, Sigmon 1974, Hamada 1985, Kumakura 1989) and/or to the iliotibial tract (Beddard 1893, Sigmon 1974, Hamada 1985). The short head inserts onto the fibular head and fascia of the leg (Champneys 1872, Hepburn 1892, Beddard 1893, Dwight 1895, Prejzner-Morawska & Urbanowick 1971, Sigmon 1974, Hamada 1985, Kumakura 1989) and occasionally onto the lateral intermuscular septum (Prejzner-Morawska & Urbanowick 1971).

- Usual innervation: Caput longum by flexores femoris nerve as is usually the case in other apes (and not by the the tibial nerve, which is usually the case in modern humans); caput breve by common fibular nerve as is usually the case in other apes and in modern humans (Champneys 1872, Sigmon 1974: *P. troglodytes*).

Semimembranosus (LSB 169.2 g; RSB 158.4 g; Fig. 44)
- Usual attachments: From the ischial tuberosity, distal and lateral to the semitendinosus (Champneys 1872, Sigmon 1974) and sometimes fused with this latter muscle (Beddard 1893, Champneys 1872), to the dorsal surface of the medial tibial condyle through a rounded tendon (Champneys 1872, Hepburn 1892, Beddard 1893, Sigmon 1974).
- Usual innervation: By the flexores femoris nerve, as is usually the case in other apes (not by the the tibial nerve, as is usually the case in modern humans, Sigmon 1974: *P. troglodytes*).

Semitendinosus (LSB 197.3 g; RSB 203.6 g; Figs. 43–45)
- Usual attachments: From the ischial tuberosity in common with the long head of the bíceps femoris (Champneys 1872, Hepburn 1892, Sigmon 1974) and sometimes also with the semimembranosus (Champneys 1872, Hepburn 1892, Beddard 1893, Sigmon 1974), to the medial tibial surface just distal to the tibial tuberosity through a narrow, flat tendon (Champneys 1872, Beddard 1893) extending further distally than in modern humans and *Gorilla* (Champneys 1872, Hepburn 1892). In some specimens of great apes there is also an aponeurotic expansion to the fascia of the leg in this region (Humphry 1867, Champneys 1872). An oblique tendinous intersection in the muscle belly of the semitendinosus is occasionally present (MacAlister 1872, Hepburn, 1892).
- Usual innervation: By the flexores femoris nerve, as is usually the case in other apes (not by the the tibial nerve, as is usually the case in modern humans; Champneys 1872, Sigmon 1974: *P. troglodytes*).

Extensor digitorum longus (LSB 77.5 g; RSB 73.0 g; Figs. 46, 48, 52)
- Usual attachments: From the head and medial crest of the fibula, intermuscular septum and the lateral tibial condyle (Beddard 1893, MacDowell 1910, Miller 1952, Kaplan 1958a, Lewis 1966, 1989) and sometimes also from the interosseus membrane (Beddard 1893, MacDowell 1910, Kaplan 1958a) and/or the crural fascia (Beddard 1893, Kaplan 1958a), to the dorsal aponeurosis of digits 2 to 4 (Hepburn 1892, Beddard 1893, MacDowell 1910, Miller 1952, Lewis 1966), the slip to digit 2 being occasionally atrophied (Beddard 1893).
- Usual innervation: Branches of the deep fibular nerve (Miller 1952: *P. paniscus*).

Extensor hallucis longus (LSB 26.1 g; RSB 24.9 g; Figs. 46, 48, 54)
- Usual attachments: Originates from the medial surface of the fibula and the interosseous membrane (Beddard 1893, Miller 1952, Lewis 1966) and

occasionally also from the lateral tibial condyle (Lewis 1966) and/or the crural fascia and intermuscular septa (Beddard 1893). The tendon of insertion passes along the shaft of MI (Hepburn 1892, Beddard 1893) to the dorsal aponeurosis of the hallux (Miller 1952) and inserts onto the distal phalanx of the hallux (Beddard 1893, Miller 1952, Lewis 1966) and occasionally also onto the proximal phalanx of this digit (Miller 1952).
- Usual innervation: Branches of the deep portion of the fibular nerve (Miller 1952: *P. paniscus*).
- Notes: According to the literature review of Gibbs (1999), the **fibularis tertius** ('peroneus tertius') is present in 5% of *Pan*; in the chimpanzees in which we examined this region in detail (VU PT1, GWUANT 1 and GWUANT 2) the fibularis tertius is missing.

Tibialis anterior (LSB 125.4 g; RSB 117.5 g; Figs. 46, 54)
- Usual attachments: From the lateral aspect of the tibia (Sutton 1883, Beddard 1893, MacDowell 1910, Miller 1952, Lewis 1966) and sometimes from the interosseous membrane (Miller 1952), the tibialis anterior splits into two bellies relatively close to its origin, and then inserts onto the plantar surface of the medial cuneiform and onto the proximal end of the MI (Humphry 1867, Sutton 1883, Hepburn 1892, Beddard 1893, MacDowell 1910, Miller 1952, Lewis 1966). The metatarsal insertion has sometimes been described as part of a separate muscle, the abductor hallucis longus (Humphry 1867, Sutton 1883).
- Usual innervation: Branch of the deep fibular nerve (Miller 1952: *P. paniscus*).

Fibularis brevis (LSB 50.6 g; RSB 41.4 g; Figs. 46, 49, 52–53)
- Usual attachments: From the distolateral fibula and from the intermuscular septa (Beddard 1893, MacDowell 1910, Miller 1952) to the tuberosity at the base of MV and the extensor digitorum tendon (Champneys 1872, Hepburn 1892, Beddard 1893, Dwight 1895, MacDowell 1910, Lewis 1966), the muscle having sometimes a double insertion (Champneys 1872) and often having a small tendon to the proximal and/or middle phalanges of digit 5 (Vrolik 1841, Champneys 1872, MacDowell 1910, Prejzner-Morawska & Urbanowick 1971). In our VU PT1 specimen the muscle has an insertion onto the base of the proximal phalanx of digit 5 (Figs. 46, 49), whereas in our GWUANT PT2 specimen it inserts onto MV.
- Usual innervation: Branches of the superficial fibular nerve (Miller 1952: *P. paniscus*).
- Synonymy: Peroneus brevis (Gibbs 1999).

Fibularis longus (LSB 95.7 g; RSB 96.4 g; Figs. 46, 49, 52–53)
- Usual attachments: From the fibular head and adjacent proximal fibula (Beddard 1893, MacDowell 1910, Miller 1952, Lewis 1966) and sometimes also from the lateral tibial condyle (Miller 1952, Lewis 1966) and/or the intermuscular septa

(Beddard 1893, Boyer 1935), passing trough a groove on the cuboid bone (Miller 1952) to insert onto the tuberosity of MI (Heburn 1892, Beddard 1893, Dwight 1895, MacDowell 1910, Miller 1952, Lewis 1966) and occasionally via a fibrous attachment to MV (Lewis 1966; our specimen GWUANT PT2: Fig. 52).
- Usual innervation: Branches of the superficial fibular nerve (Miller 1952: *P. paniscus*).
- Synonymy: Peroneus longus (Gibbs 1999).

Gastrocnemius (LSB caput laterale 137.0 and caput mediale 121.8; RSB caput laterale + caput mediale = 251.4; Figs. 46–47)
- Usual attachments: The gastrocnemius has two heads and joins with the soleus, originating from the medial and lateral femoral condyles and the capsule of the knee joint (Beddard 1893, Hepburn 1892, Frey 1913, Miller 1952). The muscle inserts onto the calcaneal tuberosity (MacAlister 1872, Beddard 1893, Miller 1952). Within the primates dissected by us, the tendo calcaneus is better defined and more substantial in *Hylobates* and *Pongo* than it is in *Gorilla* and *Pan*, but in no extant apes is it as well-defined as is usually the case in modern humans.
- Usual innervation: Tibial nerve (Miller 1952: *P. paniscus*).

Plantaris (LSB 11.3 g; RSB 11.4 g; Fig. 47)
- Usual attachments: According to the literature review of Gibbs (1999) the plantaris is absent in 40% of *Pan*, is bilaterally present in 40% of chimpanzees and unilaterally present in 20% of chimpanzees. The muscle was unilaterally present in the two chimpanzee specimens in which we analyzed this region in detail (GWUANT PT2 and VU PT1). It ran from the lateral femoral supracondylar line, the lateral femoral epicondyle and the oblique popliteal ligament to the calcaneal tuberosity.
- Usual innervation: Tibial nerve (Miller 1952: *P. paniscus*).

Soleus (LSB 224.7 g; RSB 212.0 g; Fig. 47)
- Usual attachments: It has an origin from the head and dorsal aspect of the fibular shaft (Humphry 1867, Sutton 1883, Hepburn 1892, Beddard 1893, Miller 1952, Lewis 1962), although this origin is sometimes reported as absent (Vrolik 1841). The tibial origin is as often absent (Champneys 1872, MacAlister 1873, MacDowell 1910) as present (Vrolik 1841, Humphry 1867, Champneys 1872, MacAlister 1873, Hepburn 1892, MacDowell 1910, Frey 1913, Lewis 1962), but it was present in the two chimpanzees (GWUANT PT2 and VU PT1) in which we analyzed this region in detail. The tibial head may be included with the lateral part of the gastrocnemius in those cases where it is reported as absent (Gibbs 1999). The insertion of the soleus is onto the calcaneal tuberosity (MacAlister 1873, Beddard 1893, Miller 1952).
- Usual innervation: Tibial nerve (Miller 1952: *P. paniscus*).

Flexor digitorum longus (68.1 g; RSB 66.3 g; Figs. 47–48, 50, 53)
- Usual attachments: From the dorsal aspect of the tibial shaft (Sutton 1883, Beddard 1893, MacDowell 1910, Miller 1952) and sometimes from the intermuscular septum (Beddard 1893), the tendon distribution to the digits varies among the apes. According to the literature review of Gibbs (1999) digit 5 is supplied in all apes, and in *Pan* digit 4 is supplied in 4/23, digit 3 in 2/23 specimens and digit 2 in 19/23 specimens (Beddard 1893, Champneys 1872, Dwight 1895, Hepburn 1892, Humphry 1867, Keith 1894, Lewis 1962, MacDowell 1910, Miller 1952, Sutton 1883). When present, the tendon to digit 2 is often fused with the flexor hallucis longus (Champneys 1872, Miller 1952), and the muscle may be fused to a lesser or greater degree with the tendon of flexor digitorum brevis (Sutton 1883). In our specimens GWUANT PT2 and VU PT1 the insertion is onto digits 2, 3, 4 and 5.
- Usual innervation: Tibial nerve (Miller 1952: *P. paniscus*).
- Synonymy: Flexor digitorum medialis (Gibbs 1999).

Flexor hallucis longus (LSB 136.1 g; RSB 143.6 g; Figs. 47, 50, 53)
- Usual attachments: From the interosseous membrane and posterior crural intermuscular septum and often also from the distodorsal aspect of the fibula (Sutton 1883, Champneys 1872, Beddard 1893, Miller 1952); in one case the fibular origin was reported as the proximal two-thirds (Beddard 1893). The distal attachment is to the base of the distal phalanx of the hallux and sometimes to digits 2, 3 and/or 4 (Humprhy 1867, Champneys 1872, Sutton 1883, Hepburn 1892, MacDowell 1910, Miller 1952, Lewis 1962). In our specimen VU PT1 the muscle inserts onto digits 1, 3 and 4 (Fig. 50) while in our specimen GWUANT PT2 it inserts onto digits 1, 2, 3 and 4 (Fig. 53).
- Usual innervation: Tibial nerve (Miller 1952: *P. paniscus*).
- Synonymy: Flexor digitorum lateralis (Gibbs 1999).

Popliteus (LSB 50.8 g; RSB 43.2 g; Figs. 47–48)
- Usual attachments: From the lateral femoral condylar head (Hepburn 1892, Beddard 1893, Miller 1952) and often also from the knee capsule (Hepburn 1892, Miller 1952), as well as from the tibial collateral ligament in our specimen VU PT1, to the dorsal tibial surface (Beddard 1893, Miller 1952) as well as to the lateral surface of this latter bone in our specimen VU PT1.
- Usual innervation: Tibial nerve (Miller 1952: *P. paniscus*).

Tibialis posterior (LSB 137.0 g; RSB 125.3 g; Fig. 48)
- Usual attachments: From the interosseous membrane and the adjoining surfaces of the tibia and fibula (Beddard 1893, Miller 1952) and occasionally from the intermuscular septa (Beddard 1893), the medial part of the muscle inserting onto the navicular bone (Hepburn 1892, Lewis 1964) and the lateral part onto MII, MIII and MIV (Miller 1952, Lewis 1964, onto the plantar ligaments (Hepburn 1892, Beddard 1893), and/or onto the sheath of the tendon of

fibularis longus (Humphry 1867, Hepburn 1892, Lewis, 1964). A complete cuneiform insertion is present in *P. paniscus* according to Miller (1952), while in *P. troglodytes* it can be to the medial (Humphry 1867, Lewis 1964), medial and lateral (Hepburn 1892) or lateral (Champneys 1872) cuneiforms, but never to all three.
- Usual innervation: Tibial nerve (Miller 1952: *P. paniscus*).

Extensor digitorum brevis (LSB 23.1 g; RSB 23.8 g; Figs. 49–50, 52)
- Usual attachments: The calcaneal origin and division into three tendons for the dorsal aponeurosis of digits 2, 3 and 4 is present in chimpanzees (Champneys 1872, Ruge, 1878a, Hepburn 1892, Beddard 1893, Miller 1952, Lewis 1966), and sometimes there is an additional tendon for digit 5 that merges with the tendon of the fibularis brevis.
- Usual innervation: Deep branch of the fibular nerve (Miller 1952: *P. paniscus*).

Extensor hallucis brevis (LSB 8.7 g; RSB 7.6 g; Figs. 46, 48–49)
- Usual attachments: A separate muscle that originates from the extensor digitorum brevis according to Champneys (1872), Hepburn (1892), Beddard (1893), Miller (1952) and Lewis (1966) and which according to Hepburn (1892), Miller (1952) and Lewis (1966) inserts onto the base of the proximal phalanx of the hallux, while Dwight (1895) reported an insertion onto the distal phalanx of this digit in one chimpanzee specimen. In our specimens GWUANT PT2 and VU PT1 the extensor hallucis brevis is a distinct muscle running from the calcaneus to the proximal phalanx of the hallux.
- Usual innervation: Deep branch of the fibular nerve (Miller 1952: *P. paniscus*).

Abductor digiti minimi (LSB 11.7 g; RSB 11.6 g; Figs. 50, 53–54)
- Usual attachments: From the medial and lateral calcaneus (Champneys 1872, Beddard 1893, Miller, 1952) and sometimes from the plantar aponeurosis (Champneys 1872, Beddard 1893) and/or the calcaneal tuberosity (our specimen VU PT1), to the proximal phalanx of digit 5 and the base of MV (Champneys 1872, Beddard 1893, Miller 1952) and occasionally to the distal middle phalanx of digit 5 (Vrolik 1841).
- Usual innervation: Lateral plantar nerve (Miller 1952: *P. paniscus*).
- Synonymy: Abductor digiti quinti (Gibbs 1999).

Abductor hallucis (LSB 45.0 g; RSB 36.1 g; Figs. 50, 53–54)
- Usual attachments: The abductor hallucis runs from the medial and plantar surfaces of the calcaneus (Brooks 1887, Hepburn 1892, Beddard 1893, Miller 1952) and sometimes also from the medial part of the plantar aponeurosis (Hepburn 1892, Beddard 1893), to the base of the proximal phalanx of the hallux (Vrolik 1841, Brooks 1887, Hepburn 1892, Beddard 1893, Miller 1952) and occasionally also to the medial and/or lateral cuneiforms (Vrolik 1841). According to Chapman (1879) and Miller (1952) some chimpanzees have a separate slip that goes to the base of MI, which has been named the

'abductor ossis metacarpi hallucis' ('abductor os metatarsi digiti quinti' *sensu* Gibbs 1999). In our specimen GWANT PT2 there is a slip deep to the main body of the abductor hallucis that runs from the medial surface of the calcaneus and from the navicular bone to insert together with the main tendon of the abductor hallucis onto the proximal phalanx of the hallux; this slip seems to correspond to the 'abductor ossis metacarpi hallucis' of other authors (Fig. 54).
- Usual innervation: Medial plantar nerve (Miller 1952: *P. paniscus*).

Flexor digitorum brevis (LSB 21.0 g; RSB 19.7 g; Figs. 50, 53, 54)
- Usual attachments: From the calcaneus, usually the medial and ventral surfaces (Champneys 1872, Sutton 1883, Hepburn 1892, Beddard 1893, Dwight 1895, MacDowell 1910, Miller 1952, Sarmiento 1983), and sometimes also from the proximal plantar aponeurosis (Champneys 1872) and/or the plantar aspect of the flexor digitorum longus (Sutton 1883, Hepburn 1892, MacDowell 1910, Miller 1952), to digits 2 and 3 and often to digits 4 and 5, the tendon to digit 5 being derived from a deep head that occasionally supplies digit 4 (Hepburn 1892, Sarmiento 1983). The muscle goes to digits 2, 3 and 4 in our specimen VU PT1 and to digits 2 and 3 in our specimen GWUANT PT2 (Fig. 53).
- Usual innervation: Medial plantar nerve (Miller 1952: *P. paniscus*).

Quadratus plantae (LSB 3.9 g; RSB 1.8 g; Figs. 50, 53)
- Usual attachments: From the lateral margin of the plantar surface of the calcaneus (lateral head, because the medial head is absent in *Pan*; Champneys 1872, Hepburn 1892) to the common tendon of the flexor digitorum longus before the differentiation of the flexor hallucis longus (Chapman 1879, Hepburn 1892), with the exception of some specimens of *Pan* in which the quadratus plantae was present unilaterally and did not reach the flexor digitorum longus (Humphry 1867, Champneys 1872). Attachment to the flexor digitorum longus tendons to digits 3 and 4 was reported in a specimen of *Pan* (Dwight 1895).
- Usual innervation: Data are not available.
- Synonymy: Flexor accessorius (Gibbs 1999)

Lumbricales (LSB lumbricales 1 + 2 + 3 + 4 = 6.5 g; RSB lumbricales 1 + 2 + 3 + 4 = 9.0 g; Figs. 50, 54)
- Usual attachments: There are usually four lumbricales in *Pan* (Sutton 1883, Hepburn 1892, Beddard 1893, Dwight 1895, Miller 1952; our specimens GWUANT PT2 and VU PT1) although occasionally only three muscles are present (Beddard 1893), Humphry (1867) reported seven lumbricales in *Pan*, but this is because he included two muscle structures going to the hallux. The second, third and fourth lumbricals have a double origin and insert respectively onto digits 3, 4 and 5, while the first lumbrical usually has a single head from the flexor digitorum longus tendon to digit 2 (Champneys 1872, Sutton 1883, Hepburn 1892, Beddard 1893, Dwight 1895, Miller 1952), although in

Beddard's (1893) specimen with three lumbricales the first one was reported as having a double origin, with the additional origin arising from the tendon of the flexor hallucis longus to digit 2. The tendons of the lumbricales radiate into the extensor aponeuroses of their respective digits (Sutton 1883, Miller 1952).
- Usual innervation: Medial (to lumbricales 1 and 2) and lateral (to lumbricales 3 and 4) plantar nerves (Miller 1952: *P. paniscus*).

Adductor hallucis (LSB 39.5 g; RSB 40.9 g; Figs. 46, 48–50, 53–54)
- Usual attachments: The two heads of the adductor hallucis are variably united in *Pan*; the oblique head originates from MII and MIII and the sheath of the tendon of the fibularis longus (Sutton 1883, Champneys 1872, Brooks 1887, Beddard 1893, Miller 1952) and sometimes also from MIV (Sutton 1883, Brooks 1887) and/or from the cuboid (Vrolik 1841). The transverse head originates from MII and MIII and often also from MIV (Vrolik 1841, Dwight 1895, Miller 1952), the third and fourth metatarsophalangeal joints and associated ligaments, and/or the second metatarsophalangeal joint and interosseous fascia (Brooks 1887, Beddard 1893, Miller 1952). The oblique head inserts onto the base of the proximal phalanx of the hallux (Sutton 1883, Brooks 1887) and often also to MI (Champneys 1872, Brooks 1887, Beddard 1893) and occasionally also to the distal phalanx (Brooks 1887). The insertion of the transverse head is to the base of the proximal phalanx (Sutton 1883, Brooks 1887, Miller 1952) and sometimes also to MI (Champneys 1872, Brooks 1887, Beddard 1893).
- Usual innervation: Deep branch of lateral plantar nerve and also by medial plantar nerve (Brooks 1887, Hepburn 1892, Sonntag 1923: *P. troglodytes*); deep branch of lateral plantar nerve (Miller 1952: *P. paniscus*).
- Notes: In modern humans the '**transversalis pedis**' (not listed in Terminologia Anatomica 1998) corresponds to the caput transversum of the adductor hallucis.

Flexor digiti mimini brevis (LSB present but mass undetermined; RSB 4.3 g; Fig. 51)
- Usual attachments: From the base of MV (Miller 1952), the origin being sometimes shared with the opponens digiti minimi and the plantar interossei (Champneys 1872), to the base of the proximal phalanx of digit 5 (Miller 1952). The flexor digiti minimi brevis may be absent in some *Pan* according to Champneys (1872) and is seems to be absent in our specimen GWUANT PT2, but it is present in the specimen VU PT1: Fig. 51.
- Usual innervation: Lateral plantar nerve (Miller 1952: *P. paniscus*).
- Synonymy: Flexor digiti quinti brevis or flexor digiti minimi (Gibbs 1999).

Flexor hallucis brevis (LSB present but mass undetermined; RSB caput mediale + caput laterale = 13.8 g; Figs. 50, 53–54)
- Usual attachments: The origin is usually (Brooks 1887, Hepburn 1892, Beddard 1893), but according to Lewis (1964) not always, double-headed, originating from the intermediate cuneiform and the tendon of the tibialis posterior (Vrolik 1841, Lewis 1964), the medial and lateral cuneiforms (Champneys 1872, Brooks 1887, Miller 1952) and/or the metatarsophalangeal joint (Brooks 1887) and occasionally from the cuboid (Beddard 1893). The two heads are subequal in size in *Pan* (Brooks 1887), the entire muscle being often fused with the abductor hallucis and the lateral lateral head being often fused with the adductor hallucis (Champneys 1872, Ruge 1878b, Beddard 1893). Insertion of the flexor hallucis brevis is onto the proximal phalanx of the hallux and sometimes also onto MI (Brooks 1887, Beddard 1893, Miller 1952).
- Usual innervation: Medial plantar nerve (Champneys 1872, Brooks 1887: *P. troglodytes*; Miller 1952: *P. paniscus*).

Opponens digiti minimi (LSB 1.0 g; Fig. 51)
- Usual attachments: In orangutans and modern humans the opponens digiti minimi is usually considered to be a deep fascicle of the flexor digiti minimi (for instance, it was not listed in Terminologia Anatomica 1998), but it has been described as a distinct muscle in *Pan* and *Gorilla* (e.g., Gibbs 1999). When present, it originates from the tendon sheath of the fibularis longus (Miller 1952), having a common origin from the flexor digiti minimi brevis and the fourth plantar interosseous muscle according to Champneys (1872), and inserts onto the lateral surface of MV (Champneys 1872, Miller 1952), also extending onto the plantar surface of this latter bone according to Champneys (1872). We also found an insertion onto the proximal phalanx of digit 5 in our chimpanzee specimens.
- Usual innervation: Lateral plantar nerve (Miller 1952: *P. paniscus*).
- Notes: The **opponens hallucis** has been described as a variant in *Pan* (Vrolik 1841, Champneys 1872); it was not seen in the chimpanzees in which we analyzed this region in detail (GWUANT PT2 and VU PT1).
- Synonymy: Opponens digiti quinti (Gibbs 1999).

Interossei dorsales (LSB interossei dorsales 1 + 2 + 3 + 4 = 18.8 g; RSB interossei dorsales 1 + 2 + 3 + 4 + interossei plantares 1 + 2 + 3 = 38.1; Figs. 46, 49)
- Usual attachments: There are usually four dorsal interossei muscles in *Pan* (Humphry 1867, Miller 1952), with the exception of single specimen dissected by Dwight (1895) that had six. Each muscle usually has two heads of origin. The reference line for action is through digit 3 (Champneys 1872). The first dorsal interosseous originates from the lateral surface of MI and the medial surface

of MII (Champneys 1872, Brooks 1887, Dwight 1895, Sonntag 1924, Miller, 1952), although the hallucial head may be rudimentary or absent (Brooks 1887). The second dorsal interosseous originates from the lateral surface of MII and the medial surface of MIII (Dwight 1895, Miller 1952). The third dorsal interosseous originates from the lateral surface of MIII and the medial surface of MIV (Dwight 1895, Miller 1952). The fourth dorsal interosseous originates from the lateral surface of MIV and the medial surface of MV. In the specimen of *Pan* reported by Dwight (1895) there was a first dorsal accessory inerosseous muscle lying between the third and fifth dorsal interossei muscles, originating from the dorsal surface of MIV, the second accessory dorsal interosseous muscle arose with the fourth dorsal interosseous from the lateral surface of MIV and the medial surface of MV; the fourth dorsal and second accessory dorsal interossei muscles might be considered as a single muscle with two separate insertions according to Dwight (1895). The first dorsal interosseous inserts onto the medial surface of the base of the proximal phalanx of digit 2, the second dorsal interosseous inserting onto the medial surface of digit 3, the third dorsal interosseous inserting onto the lateral surface of digit 3, and the fourth dorsal interosseous inserting onto the lateral aspect of the proximal phalanx of digit 4; the first and second accessory dorsal interossei described above insert onto the medial surface of digit 4 and onto the medial surface of digit 5, respectively (Humphry 1867, Dwight 1895, Miller 1952).
- Usual innervation: Deep ramus of the lateral plantar nerve (Miller 1952: *P. paniscus*).

Interossei plantares (LSB interossei plantares 1 + 2 + 3 = 16.5 g; RSB interossei dorsales 1+ 2 + 3 + 4 + interossei plantares 1 + 2 + 3 = 38.1; Fig. 51)
- Usual attachments: There are usually three plantar interossei muscles in *Pan*, each with a single head of origin (Humphry 1867, Miller 1952), with the exception of a specimen of *Pan* stated to possess five muscles (Dwight 1895). The first plantar interosseous usually originates from MII, the second from the medial surface of MIV, and the third from the lateral side of MIV (Dwight 1895). The two accessory plantar interossei reported by Dwight (1895) originated from the medial and lateral sides of MIII respectively. The first plantar interosseous inserts onto the lateral side of the proximal phalanx of digit 2, the second onto medial side of digit 4, and the third onto the medial side of digit 5, the two accessory plantar interossei described above inserting onto the medial and lateral sides of the proximal phalanx of digit 3 respectively (Humphry 1867, Dwight 1895, Humphry 1867, Miller 1952).
- Usual innervation: Lateral plantar nerve (Miller 1952: *P. paniscus*).

Appendix I
Literature Including Information about the Muscles of Chimpanzees

Abramowitz I (1955) On the existence of a palmar interosseous muscle in the thumb with particular reference to the Bantu-speaking Negro. *South African J Sci* 51, 270–276.
Aiello L, Dean C (1990) *An introduction to human evolutionary anatomy*. San Diego: Academic Press.
Andrews P, Groves CP (1976) Gibbons and brachiation. In *Gibbon and Siamang, Vol. 4* (ed. Rumbaugh DM), pp. 167–218. Basel: Karger.
Ashton EH, Oxnard CE (1963) The musculature of the primate shoulder. *J Zool Soc Lond* 29, 553–650.
Ashton EH & Oxnard CE (1964) Functional adaptations of the primate shoulder girdle. *Proc Zool Soc Lond* 142, 49–66.
Aversi-Ferreira TA, Diogo R, Potau JM, Bello G, Pastor JF, Aziz MA (2010) Comparative anatomical study of the forearm extensor muscles of *Cebus libidinosus* (Rylands et al. 2000; Primates, Cebidae), modern humans, and other primates, with comments on primate evolution, phylogeny and manipulatory behavior. *Anat Rec* 293, 2056–2070.
Avril C (1963) Kehlkopf und Kehlsack des Schimpansen (*Pan troglodytes*). *Gegen Morphol Jahrb* 105, 74–129.
Aziz MA, Dunlap SS (1986) The human extensor digitorum profundus muscle with comments on the evolution of the primate hand. *Primates* 27, 293–319.
Barnard WS (1875) Observations on the membral musculation of *Simia satyrus* (Orang) and the comparative myology of man and the apes. *Proc Amer Assoc Adv Sci* 24, 112–144.
Beddard FE (1893) Contributions to the anatomy of the anthropoid apes. *Trans Zool Soc Lond* 13, 177–218.
Bello-Hellegouarch G, Potau JM, Arias-Martorell J, Diogo R, Perez-Perez A (2012). The rotator cuff muscles in Hominoidea: evolution and adaptations to different types of locomotion. In *Primates: Classification, evolution and behavior* (eds. Hughes EF, Hill ME). Hauppauge, Nova Science Publishers.
Blake ML (1976) *The quantitative myology of the hind limb of Primates with special reference to their locomotor adaptations*. Unpublished PhD Thesis, University of Cambridge, Cambridge.
Blair DM (1923) A study of the central tendon of the diaphragm. *J Anat Phys* 57, 203–215.
Bijvoet WF (1908) Zur vergleichenden Morphologie des Musculus digastricus mandibulae bai den Säugetieren. *Zeit Morphol Anthropol* 11, 249–316.

Bischoff TLW (1870) Beitrage zur Anatomie des *Hylobates leuciscus* and zueiner vergleichenden Anatomie der Muskeln der Affen und des Menschen. *Abh Bayer Akad Wiss München Math Phys Kl* 10, 197–297.

Bojsen-Møller F (1978) Extensor carpi radialis longus muscle and the evolution of the first intermetacarpal ligament. *Am J Phys Anthropol* 48, 177–184.

Bolk L (1902) Beiträge zur Affenanatomie, III, Der Plexus cervico-brachialis der Primaten. *Petrus Campter* 1, 371–566.

Boyer EL (1935) The musculature of the inferior extremity of the orang-utan, *Simia satyrus*. *Am J Anat* 56, 192–256.

Broca P (1869) L'ordre des primates—parallele anatomique de l'homme et des singes. *Bull Soc Anthropol Paris* 4, 228–401.

Brooks HSJ (1886a) On the morphology of the intrinsic muscles of the little finger, with some observations on the ulnar head of the short flexor of the thumb. *J Anat Physiol* 20, 644–661.

Brooks HSJ (1886b) Variations in the nerve supply of the flexor brevis pollicis muscle. *J Anat Physiol* 20, 641–644.

Brooks HSJ (1887) On the short muscles of the pollex and hallux of the anthropoid apes, with special reference to the opponens hallucis. *J Anat Physiol* 22, 78–95.

Brown B (1983) An evaluation of primate caudal musculature in the identification of the ischiofemoralis muscle. *Am J Phys Antropol* 60, 177–178.

Burrows AM (2008) The facial expression musculature in primates and its evolutionary significance. *Bioessays* 30, 212–225.

Burrows AM, Waller BM, Parr LA, Bonar CJ (2006) Muscles of facial expression in the chimpanzee (*Pan troglodytes*): descriptive, comparative and phylogenetic contexts. *J Anat* 208, 153–167.

Burrows AM, Waller BM, Parr LA (2009) Facial musculature in the rhesus macaque (*Macaca mulatta*): evolutionary and functional contexts with comparisons to chimpanzees and humans. *J Anat* 215, 320–334.

Carlson KJ (2006) Muscle architecture of the common chimpanzee (*Pan troglodytes*): perspectives for investigating chimpanzee behavior. *Primates* 47, 218–229.

Cave AJE (1979) The mammalian temporo-pterygoid ligament. *J Zool Lond* 188, 517–532.

Champneys F (1872) The muscles and nerves of a Chimpanzee (*Troglodytes Niger*) and a *Cynocephalus Anubis*. *J Anat Physiol* 6, 176–211.

Chapman HC (1879) On the structure of the chimpanzee. *Proc Acad Nat Sci Philad* 31, 52–63.

Cihak R (1972) Ontogenesis of the skeleton and intrinsic muscles of the human hand and foot. *Adv Anat Embryol Cell Biol* 46, 1–194.

Clegg M (2001) *The comparative anatomy and evolution of the human vocal tract.* Unpublished PhD Thesis, University of London, London.

Day MH, Napier J (1963) The functional significance of the deep head of flexor pollicis brevis in primates. *Folia Primatol* 1, 122–134.

Dean MC (1984) Comparative myology of the hominoid cranial base, I, the muscular relationships and bony attachments of the digastric muscle. *Folia Primatol* 43, 234–48.

Dean MC (1985) Comparative myology of the hominoid cranial base, II, the muscles of the prevertebral and upper pharyngeal region. *Folia Primatol* 44, 40–51.

Diogo R, Abdala V (2010) Muscles of vertebrates—comparative anatomy, evolution, homologies and development. Oxford: Taylor & Francis.

Diogo R, Wood BA (2008) Comparative anatomy, phylogeny and evolution of the head and neck musculature of hominids: a new insight. *Am J Phys Anthropol, Suppl* 46, 90.

Diogo R, Wood BA (2009) Comparative anatomy and evolution of the pectoral and forelimb musculature of primates: a new insight. *Am J Phys Anthropol, Meeting Suppl* 48, 119.

Diogo R, Wood BA (2011) Soft-tissue anatomy of the primates: phylogenetic analyses based on the muscles of the head, neck, pectoral region and upper limb, with notes on the evolution of these muscles. *J Anat* 219, 273–359.

Diogo R, Wood BA (2012) Comparative anatomy and phylogeny of primate muscles and human evolution. Oxford: Taylor & Francis.

Diogo R, Abdala V, Lonergan N, Wood BA (2008) From fish to modern humans—comparative anatomy, homologies and evolution of the head and neck musculature. *J Anat* 213, 391–424.

Diogo R, Abdala V, Aziz MA, Lonergan N, Wood BA (2009a) From fish to modern humans—comparative anatomy, homologies and evolution of the pectoral and forelimb musculature. *J Anat* 214, 694–716.

Diogo R, Wood BA, Aziz MA, Burrows A (2009b) On the origin, homologies and evolution of primate facial muscles, with a particular focus on hominoids and a suggested unifying nomenclature for the facial muscles of the Mammalia. *J Anat* 215, 300–319.

Diogo R, Richmond BG, Wood B (in press). Evolution and homologies of modern human hand and forearm muscles, with notes on thumb movements and tool use. *J Hum Evol*.

DuBrul EL (1958) *Evolution of the speech apparatus*. Springfield: Thomas.

Duchin LE (1990) The evolution of human speech: comparative anatomy of the oral cavity in *Pan* and *Homo*. *J Hum Evol* 19, 687–697.

Duckworth WLH (1904) *Studies from the Anthropological Laboratory, the Anatomy School*, Cambridge. London: C. J. Clay & Sons.

Duckworth WLH (1912) On some points in the anatomy of the plica vocalis. *J Anat Physiol* 47, 80–115.

Duckworth WLH (1915) *Morphology and anthropology (2nd ed.)*. Cambridge: Cambridge University Press.

Dunlap SS, Thorington RW, Aziz MA (1985) Forelimb anatomy of New World monkeys: myology and the interpretation of primitive anthropoid models. *Am J Phys Anthropol* 68, 499–517.

Duvernoy M (1855-1856) Des caracteres anatomiques de grands singes pseudoanthropomorphes anthropomorphes. *Arch Mus Natl Hist Nat Paris* 8, 1–248.

Dylevsky I (1967) Contribution to the ontogenesis of the flexor digitorum superficialis and the flexor digitorum profundus in man. *Folia Morphol (Praha)* 15, 330–335.

Dwight T (1895) Notes on the dissection and brain of the chimpanzee 'Gumbo'. *Mem Boston Soc Nat Hist* 5, 31–51.

Edgeworth FH (1935) *The cranial muscles of vertebrates*. Cambridge: Cambridge University Press.

Ehlers E (1881) Beiträge zur Kenntnis des gorilla und chimpanse. *Abh d K Gesell d Wissensch du Göttingen, Phys Cl* 38, 3–77.

Elftman HO (1932) The evolution of the pelvic floor of primates. *Am J Anat* 51, 307–346.

Ferrero EM, Pastor JF, Fernandez FP, Barbosa M, Diogo R, Wood B (2012). Comparative anatomy of the lower limb muscles of hominoids: attachments, relative weights, innervation and functional morphology. In *Primates: Classification, evolution and behavior* (eds. Hughes EF, Hill ME). Hauppauge, Nova Science Publishers.

Falk D (1993) Meningeal arterial patterns in great apes: implications for hominid vascular evolution. *Am J Phys Anthropol* 92, 81–97.

Falk D, Nicholls P (1992) Meningeal arteries in Rhesus macaques (*Macaca mulatta*): implications for vascular evolution in anthropoids. *Am J Phys Anthropol* 89, 299–308.

Fick R (1895a) Vergleichend-anatomische Studien an einem erwachsenen Orang-utang. *Arch Anat Physiol Anat Abt* 1895, 1–100.

Fick R (1895b) Beobachtungen an einem zweiten envachsenen Orang Utang und einem Schimpansen. *Arch Anat Physiol Anat Abt* 1895, 289–318.

Fick R (1925) Beobachtungen an den muskeln einger schimpansen. *Zeitschr Anat und Entwick* 76, 117–141.

Fitzwilliams DCL (1910) The short muscles of the hand of the agile gibbon (*Hylobatis agilis*), with comments on the morphological position and function of the short muscles of the hand of man. *Proc R Soc Edinb* 30, 201–218.

Fleagle JG (1999) *Primate Adaptation and Evolution (2nd Ed.)*. San Diego: Academic Press.

Fleagle JG, Stern JT, Jungers WL, Susman RL, Vangor AK, Wells JP (1981) Climbing: A biomechanical link with brachiation and bipedalism. *Symp zool Soc Lond* 48, 359–373.

Forster A (1903) Die Insertion des Musculus semimembranosus. *Arch Anat Physiol Anat Abt* 1953, 257–320.

Forster A (1917) Die mm. contrahentes und interossei manus in der Säugetierreihe und beim Menschen. *Arch Anat Physiol Anat Abt* 1916, 101–378.

Frey H (1913) Der Musculus triceps surae in der Primatenreihe. *Morph Jahrb* 47, 1–192.

Giacomini C (1897) 'Plica semilunaris' et larynx chez les singes anthropomorphes. *Arch Ital Biol* 28, 98–119.

Gibbs S (1999) *Comparative soft tissue morphology of the extant Hominoidea, including Man.* Unpublished PhD Thesis, The University of Liverpool, Liverpool.

Gibbs S, Collard M, Wood BA (2000) Soft-tissue characters in higher primate phylogenetics. *Proc Natl Acad Sci US* 97, 11130–11132.

Gibbs S, Collard M, Wood BA (2002) Soft-tissue anatomy of the extant hominoids: a review and phylogenetic analysis. *J Anat* 200, 3–49.

Glidden EM, De Garis CF (1936) Arteries of the chimpanzee (*Pan* spec. ?), I, the aortic arch, II, arteries of the upper extremity, III, the descending aorta, IV, arteries of the lower extremity. *Am J Anat* 58, 501–527.

Göllner K (1982) Untersuchungen über die vom N. trigeminus innervierte Kiefermusculatur des Schimpansen (*Pan troglodytes*, Blumenbach 1799) und des Gorilla (*Gorilla gorilla gorilla*, Savage and Wyman 1847). *Gegenb Morph Jahrb* 128, 851–903.

Gratiolet LP, Alix PHE (1866) Recherches sur l'anatomie du *Troglodytes aubryi*. *Nouv Arch Mus Hist Nat Paris* 2, 1–264.

Grönroos H (1903) Die musculi biceps brachii und latissimocondyloideus bei der affengattung *Hylobates* im vergleich mit den ensprechenden gebilden der anthropoiden und des menschen. *Abh Kön Preuss Akad Wiss Berlin* 1903, 1–102.

Groves CP (1986) Systematics of the great apes. In *Comparative Primate Biology: Systematics, Evolution and Anatomy, Vol. 1* (eds. Swindler DR, Erwin J), pp. 187–217. New York: A.R. Liss.

Groves CP (1995) *Revised character descriptions for Hominoidea.* Typescript, 9 pp.

Hamada Y (1985) Primate hip and thigh muscles: comparative anatomy and dry weights. In *Primate Morphophysiology, Locomotor Analyses and Human Bipedalism* (ed. Kondo S), pp. 131–152. Tokyo: University of Tokyo Press.

Hänel H (1932) Über die Gesichtsmuskulatur der katarrhinen Affen. *Gegenbaur Morph Jahrb* 71, 1–76.

Harrison DFN (1995) *The anatomy and physiology of the mammalian larynx.* Cambridge: University Press.

Hartmann R (1886) *Anthropoid apes.* London: Keegan.

Hepburn D (1892) The comparative anatomy of the muscles and nerves of the superior and inferior extremities of the anthropoid apes: I—Myology of the superior extremity. *J Anat Physiol* 26, 149–186.

Hepburn D (1896) A revised description of the dorsal interosseous muscles of the human hand, with suggestions for a new nomenclature of the palmar interosseous muscles and some observations on the corresponding muscles in the anthropoid apes. *Trans R Soc Edin* 38, 557–565.

Herring SW, Herring SE (1974) The superficial masseter and gape in mammals. *Am Nat* 108, 56 1–576.

Hill WCO, Harrison-Matthews L (1949) The male external genitalia of the gorilla, with remarks on the os penis of other Hominoidea. *Proc Zool Soc Lond* 119, 363–378.

Himmelreich HA (1971) M. levator veli palatini der Säugetiere. *Gegen Morphol Jahrb* 116, 377–400.

Himmelreich HA (1977) M. cephalopharyngeus der Säugetiere. *Gegen Morphol Jahrb* 123, 556–588.

Hofër W (1892) Vergleichend-anatomische Studien uber die Nerven des Armes und der Hand bei den Affen und dem Menschen. *Munchener Med Abhandl* 30, 1–106.

Howell AB (1936a) Phylogeny of the distal musculature of the pectoral appendage. *J Morphol* 60, 287–315.
Howell AB (1936b) The phylogenetic arrangement of the muscular system. *Anat Rec* 66, 295–316.
Howell AB, Straus WL (1932) The brachial flexor muscles in primates. *Proc US Natl Mus* 80, 1–31.
Huber E (1930a) Evolution of facial musculature and cutaneous field of trigeminus—Part I. *Q Rev Biol* 5, 133–188.
Huber E (1930b) Evolution of facial musculature and cutaneous field of trigeminus—Part II. *Q Rev Biol* 5, 389–437.
Huber E (1931) *Evolution of facial musculature and expression*. Baltimore: The Johns Hopkins University Press.
Humphry G (1867) On some points in the anatomy of the chimpanzee. *J Anat Physiol* 1, 254–268.
Huntington GS (1903) Present problems of myological research and the significance and classification of muscular variations. *Am J Anat* 2, 157–175.
Huxley TH (1864) The structure and classification of the Mammalia. *Med Times Gazette* 1864, 398–468.
Huxley TH (1871) *The anatomy of vertebrated animals*. London; J. & A. Churchill.
Imparati E (1895-1896) Contribuzione alia miologia delle regione antero-laterale del torace, costale, e della spalla, nelle Seimmie. *Eiv Ital Sci Nat Siena* 15, 118–121, 129–132, 145–148;16, 7–9, 17–24.
Jordan J (1971a) Studies on the structure of the organ of voice and vocalization in the chimpanzee, Part 1. *Folia Morphol* 30, 99–117.
Jordan J (1971b) Studies on the structure of the organ of voice and vocalization in the chimpanzee, Part 2. *Folia Morphol* 30, 222–248.
Jordan J (1971c) Studies on the structure of the organ of voice and vocalization in the chimpanzee, Part 3. *Folia Morphol* 30, 323–340.
Jouffroy FK (1971) Musculature des membres. In *Traité de Zoologie, XVI: 3 (Mammifères)* (ed. Grassé PP), pp. 1–475. Paris: Masson et Cie.
Jouffroy FK, Lessertisseur J (1959) Reflexions sur les muscles contracteurs des doigts et des orteils (contrahentes digitorum) chez les primates. *Ann Sci Nat Zool, Ser 12*, 1, 211–235.
Jouffroy FK, Lessertisseur J (1960) Les spécialisations anatomiques de la main chez les singes à progression suspendue. *Mammalia* 24, 93–151.
Jouffroy FK, Saban R (1971) Musculature peaucière. In *Traité de Zoologie, XVI: 3 (Mammifères)* (ed. Grassé PP), pp. 477–611. Paris: Masson et Cie.
Juraniec J (1972) The aortic and esophageral hiatus in the diaphragm of primates. *Folia Morphol* 31, 197–207.
Juraniec J, Szostakiewicz-Sawicka H (1968) The central tendon of the diaphragm in primates. *Folia Morphol* 27, 183–194.
Kaneff A (1959) Über die evolution des m. abductor pollicis longus und m. extensor pollicis brevis. *Mateil morphol Inst Bulg Akad Wiss* 3, 175–196.
Kaneff A (1968) Zur differenzierung des m. abductor pollicis biventer beim Menschen. *Gegenbaurs morphol Jahrb* 112, 289–303.
Kaneff A (1969) Umbildung der dorsalen Daumenmuskeln beim Menschen. *Verh Anat Ges* 63, 625–636.
Kaneff A (1979) Évolution morphologique des musculi extensores digitorum et abductor pollicis longus chez l'Homme. I. Introduction, méthodologie, M. extensor digitorum. *Gegenbaurs Morphol Jahrb* 125, 818–873.
Kaneff A (1980a) Évolution morphologique des musculi extensores digitorum et abductor pollicis longus chez l'Homme. II. Évolution morphologique des m. extensor digiti minimi, abductor pollicis longus, extensor pollicis brevis et extensor pollicis longus chez l'homme. *Gegenbaurs Morphol Jahrb* 126, 594–630.

Kaneff A (1980b) Évolution morphologique des musculi extensores digitorum et abductor pollicis longus chez l'Homme. III. Évolution morphologique du m. extensor indicis chez l'homme, conclusion générale sur l 'évolution morphologique des musculi extensores digitorum et abductor pollicis longus chez l'homme. *Gegenbaurs Morphol Jahrb* 126, 774–815.

Kaneff A (1986) Die Aufrichtung des Menschen und die mor phologisches Evolution der Musculi extensores digitorum pedis unter dem Gesichtpunkt der evolutiven Myologie, Teil I. *Morph Jahrb* 132, 375–419.

Kaneff A, Cihak R (1970) Modifications in the musculus extensor digitorum lateralis in phylogenesis and in human ontogenesis. *Acta Anat Basel* 77, 583–604.

Kaplan EB (1958a) The iliotibial tract—clinical and morphological significance. *J Bone Jt Surg* 40A, 817–832.

Kaplan EB (1958b) Comparative anatomy of the extensor digitorum longus in relation to the knee joint. *Anat Rec* 131, 129–149.

Kaseda M, Nakamura M, Ischihara N, Hayakawa T and Asari M (2008) A macroscopic examination of M. Biceps Femoris and M. Gluteus Maximus in the Orangutan. *J Vet Med Sci* 70, 217–222.

Keith A (1894a) *The myology of the Catarrhini: a study in evolution.* Unpublished PhD thesis, University of Alberdeen, Alberdeen.

Keith A (1894b) Notes on a theory to account for the various arrangements of the flexor profundus digitorum in the hand and foot of primates. *J Anat Physiol* 28, 335–339.

Keith A (1896) A variation that occurs in the manubrium stemi of higher primates. *J Anat Phys* 30, 275–279.

Keith A (1899) On the chimpanzees and their relationship to the gorilla. *Proc Zool Soc Lond* 1899, 296–312.

Kelemen G (1948) The anatomical basis of phonation in the chimpanzee. *J Morphol* 82, 229–256.

Kelemen G (1969) Anatomy of the larynx and the anatomical basis of vocal performance. In *The Chimpanzee—Vol 1—Anatomy, behaviour and diseases of chimpanzees* (ed. Bourne GH), pp. 35–186. Basel: Karger.

Kikuchi Y (2010a) Comparative analysis of muscle architecture in primate arm and forearm. *Anat Histol Embryol* 39, 93–106.

Kikuchi Y (2010b) Quantitative analysis of variation in muscle internal parameters in crab-eating macaques (*Macaca fascicularis*). *Anthropol Sci* 118, 9–21.

Kleinschmidt A (1950) Zur anatomie des kehlkopfs der Anthropoiden. *Anat Anz* 97, 367–372.

Kohlbrügge JHF (1896) Der larynx und die stimmbildung der Quadrumana. *Natuurk T Ned Ind* 55, 157–175.

Kohlbrügge JHF (1897) Muskeln und Periphere Nerven der Primaten, mit besonderer Berücksichtigung ihrer Anomalien. *Verh K Akad Wet Amsterdam Sec 2* 5, 1–246.

Koizumi M, Sakai T (1995) The nerve supply to coracobrachialis in apes. *J Anat* 186, 395–403.

Körner F (1929) Vergleichend-anatomische Untersuchungen über den Faserverlauf der Pars lumbalis des Zwerchfells zur Begrenzung des Hiatus oesophageus bei Säugetieren. *Morphol Jahrb* 61, 409–451.

Kumakura H (1989) Functional analysis of the biceps femoris muscle during locomotor behavior in some primates. *Am J Phys Anthropol* 79, 379–391.

Körner O (1884) Beiträge zur vergleichenden Anatomie und Physiologie des Kehlkopfes der Säugethiere und des Menschen. *Abh Senckenb Naturforsch Ges* 13, 147–261.

Laitman JT (1977) *The ontogenetic and phylogenetic development of the upper respiratory system and basicranium in man.* Unpublished PhD thesis, Yale University, New Haven.

Lander KF (1918) The pectoralis minor: a morphological study. *J Anat* 52, 292–318.

Landsmeer JM (1984) The human hand in phylogenetic perspective. *Bull Hosp Jt Dis Orthop Inst* 44, 276–287.

Landsmeer JM (1986) A comparison of fingers and hand in *Varanus*, opossum and primates. *Acta Morphol Neerl Scand* 24, 193–221.

Landsmeer JM (1987) The hand and hominisation. *Acta Morphol Neerl Scand* 25, 83–93.

Larson SG, Stern JT (1986) EMG of scapulohumeral muscles in the chimpanzee during reaching and 'arboreal' locomotion. *Am J Anat* 176, 171–190.

Larson SG, Stern JT, Jungers WL (1991) EMG of serratus anterior and trapezius in the chimpanzee: scapular rotators revisited. *Am J Phys Anthropol* 85, 71–84.

Lewis OJ (1962) The comparative morphology of M. flexor accessorius and the associated long flexor tendons. *J Anat* 96, 321–333.

Lewis OJ (1964) The evolution of the long flexor muscles of the leg and foot. In *International Review of General and Experimental Zoology* (eds. Felts WJL, Harrison RJ), pp. 165–185. New York: Academic Press.

Lewis OJ (1965) The evolution of the Mm. interossei in the primate hand. *Anat Rec* 153, 275–287.

Lewis OJ (1966) The phylogeny of the cruropedal extensor musculature with special reference to the primates. *J Anat* 100, 865–880.

Lewis OJ (1989) *Functional morphology of the evolving hand and foot*. Oxford: Clarendon Press.

Lightoller GHS (1925) Facial muscles—the modiolus and muscles surrounding the rima oris with some remarks about the panniculus adiposus. *J Anat Physiol* 60, 1–85.

Lightoller GS (1928) The facial muscles of three orang utans and two cercopithecidae. *J Anat* 63, 19–81.

Lightoller GS (1934) The facial musculature of some lesser primates and a *Tupaia*. *Proc Zool Soc Lond* 1934, 259–309.

Lightoller GS (1939) V. Probable homologues. A study of the comparative anatomy of the mandibular and hyoid arches and their musculature—Part I. Comparative myology. *Trans Zool Soc Lond* 24, 349–382.

Lightoller GS (1940a) The comparative morphology of the platysma: a comparative study of the sphincter colli profundus and the trachelo-platysma. *J Anat* 74, 390–396.

Lightoller GS (1940b) The comparative morphology of the m. caninus. *J Anat* 74, 397–402.

Lightoller GS (1942) Matrices of the facialis musculature: homologization of the musculature in monotremes with that of marsupials and placentals. *J Anat* 76, 258–269.

Loth E (1912) Beiträge zur Anthropologie der Negerweichteile (Muskelsystem). *Stud Forsch Menschen-u Völkerkunde Stuttgart* 9, 1–254.

Loth E (1931) *Anthropologie des parties molles (muscles, intestins, vaisseaux, nerfs peripheriques)*. Paris: Mianowski-Masson et Cie.

Low A (1907) A note on the crura of the diaphragm and the muscle of Treitz. *J Anat* 42, 93–96.

Lunn HF (1948) The comparative anatomy of the inguinal ligament. *Anat Phys* 82, 58–67.

Lunn HF (1949) Observations on the mammalian inguinal region. *Proc Zool Soc* 118, 345–355.

Macalister A (1871) On some points in the myology of the chimpanzee and others of the primates. *Ann Mag Nat Hist* 7, 341–351.

MacDowell EC (1910) Notes on the myology of *Anthropopithecus niger* and *Papio-thoth ibeanus*. *Am J Anat* 10, 431–460.

Maier W (2008) Epitensoric position of the chorda tympani in Anthropoidea: a new synapomorphic character, with remarks on the fissura glaseri in Primates. In *Mammalian Evolutionary Morphology: a Tribute to Frederick S. Szalay* (eds. Sargis EJ, Dagosto M), pp. 339–352. Dordrecht: Springer.

Mangini U (1960) Flexor pollicis longus muscle: its morphology and clinical significance. *J Bone Jt Surg* 42A, 467–559.

Manners-Smith T (1908) A study of the cuboid and os peroneum in the primate foot. *J Anat Phys* 42, 397–414.

Mayer JC (1856) Zur Anatomie des Orang-utang und des Schimpansen. *Arch Naturgesch* 22, 279–304.

McMurrich JP (1903a) The phylogeny of the forearm flexors. *Amer J Anat* 2, 177–209.

McMurrich JP (1903b) The phylogeny of the palmar musculature. *Amer J Anat* 2, 463–500.

Mijsberg WA (1923) Über den Bau des Urogenitalapparates bei den männlichen Primaten. *Verh K Akad Wet Amsterdam* 23, 1–92.
Miller RA (1932) Evolution of the pectoral girdle and forelimb in the primates. *Amer J Phys Anthropol* 17, 1–56.
Miller RA (1934) Comparative studies upon the morphology and distribution of the brachial plexus. *Am J Anat* 54, 143–175.
Miller RA (1945) The ischial callosities of primates. *Am J Anat* 76, 67–87.
Miller RA (1947) The inguinal canal of primates. *Am J Anat* 90, 117–142.
Miller RA (1952) The musculature of *Pan paniscus*. *Am J Anat* 91, 182–232.
Morton DJ (1922) Evolution of the human foot, part 1. *Am J Phys Anthropol* 5, 305–336.
Mustafa MA (2006) Neuroanatomy. *10th National Congress of Anatomy Bordum, Turkey, September 5*, 6–10.
Mysberg WA (1917) Über die Verbinderungen zwischen dem Sitzbeine und der Wirbelsäule bei den Säugetieren. *Anat Hefte* 54, 641–668.
Negus VE (1949) *The comparative anatomy and physiology of the larynx*. New York: Hafner Publishing Company.
Ogihara N, Kunai T, Nakatsukasa M (2005) Muscle dimensions in the chimpanzee hand. *Primates* 46, 275–280.
Oishi M, Ogihara N, Endo H, Ichihara N, Asari M (2009) Dimensions of forelimb muscles in orangutans and chimpanzees. *J Anat* 215, 373–382.
Owen R (1830-1831) On the anatomy of the orangutan (*Simia satyrus*, L.). *Proc Zool Soc Lond* 1, 4–5, 9–10, 28–29, 66–72.
Owen R (1868) *The Anatomy of Vertebrates, Vol. 3: Mammals*. London: Longmans, Green & Co.
Parsons FG (1898a) The muscles of mammals, with special relation to human myology, Lecture 1, The skin muscles and muscles of the head and neck. *J Anat Physiol* 32:428–450.
Parsons FG (1898b) The muscles of mammals, with special relation to human myology: a course of lectures delivered at the Royal College of Surgeons of England—lecture II, the muscles of the shoulder and forelimb. *J Anat Physiol* 32, 721–752.
Payne RC (2001) *Musculoskeletal adaptations for climbing in hominoids and their role as exaptations for the acquisition of bipedalism*. Unpublished PhD thesis, The University of Liverpool, Liverpool.
Pearson K, Davin AG (1921) On the sesamoids of the knee joint. *Biometrika* 13, 133–175, 350–400.
Pellatt A (1979) The facial muscles of three African primates contrasted with those of *Papio ursinus*. *S Afr J Sci* 75, 436–440.
Plattner F (1923) Über die ventral-innervierte und die genuine rückenmuskulatur bei drei anthropomorphen (*Gorilla gina*, *Hylobates* und *Troglodytes niger*). *Morphol Jb* 52, 241–280.
Potau JM, Bardina X, Ciurana N, Camprubi D, Pastor JF, De Paz F, Barbosa M (2009) Quantitative analysis of the deltoid and rotator cuff muscles in humans and great Apes. *Int J Primatol* 30, 697–708.
Potau JM, Artells R, Bello G, Muñoz C, Monzó M, Pastor JF, de Paz F, Barbosa M, Diogo R, Wood B (in press) Expression of myosin heavy chain isoforms in the supraspinatus muscle of different primate species. *Int J Prim* 32, 931–944.
Prejzner-Morawska A, Urbanowicz M (1971) The biceps femoris muscle in lemurs and monkeys. *Folia Morphol* 30, 9465–482.
Primrose A (1899) The anatomy of the orang-outang (*Simia satyrus*), an account of some of its external characteristics, and the myology of the extremities. *Trans Royal Can Inst* 6, 507–594.
Prioton JB, Colin R, Baumel H (1957) Muscles moteurs des doigts et leur innervation chez le chimpanzé. *C R Assoc Anat* XLIII, 687–693.
Ranke K (1897) Muskel-und Nervenvariationen der dorsalen elemente des Plexus ischiadicus der Primaten. *Arch Anthropol* 24, 117–144.
Rauwerdink GP (1993) Muscle fibre and tendon lengths in primate extremities. In *Hands of Primates* (eds. Preuschoft H, Chivers DJ), pp. 207–223. New York: Springer-Verlag.

Rex H (1887) Ein Beitrag zur Kenntnis der Muskulatur der Mundspalte der Affen. *Morphol Jahrb* 12:275–286.
Ribbing L, Hermansson K (1912) Kleinere muskelstudien, I11, die distale extremitatenmuskulatur eines schimpansen. *Lunds Universitets Arsskrift NF Afd 2* 8, 1–10.
Roberts JA, Seibold HR (1971) The histology of the primate urethra. *Folia Primatol* 14, 59–69.
Robinson JT. Freedman L, Sigmon BA (1972) Some aspects of pongid and hominid bipedality. *J Hum Evol* 1, 361–369.
Rogers CR, Mooney MP, Smith TD, Weinberg SM, Waller BW, Parr LA, Docherty BA, Bonar CJ, Reinholt LE, Deleyiannis FW-B, Siegel MI, Marazita ML, Burrows AM (2008) Comparative microanatomy of the orbicularis oris muscle between chimpanzees and humans: evolutionary divergence of lip function. *J Anat* 214, 36–44.
Ruge G (1878a) Untersuchung uber die Extensorengruppe aus Unterschenkel und Füsse der Säugethiere. *Morphol Jahrb* 4, 592–643.
Ruge G (1878b) Zur vergleichenden Anatomie der tiefen Muskeln in der Fussohle. *Morphol Jahrb* 4, 644–659.
Ruge G (1885) Über die Gesichtsmuskulatur der halbaffen. *Gegen Morph Jahrb* 11, 243–315.
Ruge G (1887a) *Untersuchungen uber die Gesichtsmuskeln der Primaten*. Leipzig: W. Engelmann.
Ruge G (1887b) Die vom Facialis innervirten Muskeln des Halses, Nackens und des Schädels einen jungen *Gorilla*. *Gegenb Morph Jahrb* 12, 459–529.
Ruge G (1890-1891) Anatomisches über den Rumpf der Hylobatiden—ein Beitrag zur Bestimmung der Stellung dieses genus im System. In *Zoologische Ergebnisse Einer Reise in Niederländisch Ost-Indien, Vol. 1* (ed. Weber M), pp. 366–460. Leiden: Verlag von EJ Brill.
Ruge G (1897) *Über das peripherische gebiet des nervus* facialis *boi wirbelthieren*. Leipzig: Festschr f Gegenbaur.
Ruge G (1911) Gesichtsmuskulatur und Nervus facialis der Gattung *Hylobates*. *Morph Jahrb* 44, 129—177.
Saban R (1968) Musculature de la tête. In Traité de Zoologie, XVI: 3 (Mammifères) (ed. Grassé PP), pp. 229–472. Paris: Masson et Cie.
Sakka M (1973) Anatomie comparée de l'écaille de l'occipital (squama occipitalis P.N.A.) et des muscles de la nuque chez l'homme et les pongidés, II Partie, Myologie. *Mammalia* 37, 126–180.
Sarmiento EE (1983) The significance of the heel process in anthropoids. *Int J Primatol* 4, 127–152.
Schreiber HV (1934) Zur morphologie der primatenhand—rontnenolonische untersuchungen an der handwuriel dei affen. *Anat Anz* 78, 369–429.
Schreiber HV (1936) Die extrembewegungen der schimpansenhand, 2, mitteilung zu—zur morphologie der primatehand. *Morph Jahrb* 77, 22–60.
Schück AC (1913a) Beiträge zur Myologie der Primaten, I—der m. lat. dorsi und der m. latissimo-tricipitalis. *Morphol Jahrb* 45, 267–294.
Schück AC (1913b) Beiträge zur Myologie der Primaten, II—1 die gruppe sterno-cleido-mastoideus, trapezius, omo-cervicalis, 2 die gruppe levator scapulae, rhomboides, serratus anticus. *Morphol Jahrb* 46, 355–418.
Schultz AH (1936) Characters common to higher primates and characters specific for man. *Q Rev Biol* 11, 259–283, 425–455.
Schultz AH (1973) The skeleton of the Hylobatidae and other observations on their morphology. In *Gibbon and Siamang, Vol. 2* (ed. Rumbaugh DM), pp. 1–53. Basel: Karger
Seiler R (1970) Differences in the facial musculature of the nasal and upper-lip region in catarrhine primates and man. *Z Morphol Anthropol* 62, 267–275.
Seiler R (1971a) A comparison between the facial muscles of Catarrhini with long and short muzzles. *Proc 3rd Int Congr Primat Zürich 1970, vol l, Basel: Karger*, 157–162.
Seiler R (1971b) Facial musculature and its influence on the facial bones of catarrhine Primates, I. *Morphol Jahrb* 116, 122–142.

Seiler R (1971c) Facial musculature and its influence on the facial bones of catarrhine Primates, II. *Morphol Jahrb* 116, 147–185.

Seiler R (1971d) Facial musculature and its influence on the facial bones of catarrhine Primates, III. *Morphol Jahrb* 116, 347–376.

Seiler R (1971e) Facial musculature and its influence on the facial bones of catarrhine Primates, IV. *Morphol Jahrb* 116, 456–481.

Seiler R (1973) On the function of facial muscles in different behavioral situations—a study based on muscle morphology and electromyography. *Am J Phys Anthropol* 38, 567–71.

Seiler R (1974a) Muscles of the external ear and their function in man, chimpanzees and *Macaca*. *Morphol Jahrb* 120, 78–122.

Seiler R (1976) Die Gesichtsmuskeln. In *Primatologia, Handbuch der Primatenkunde, Bd. 4, Lieferung 6* (eds. Hofer H, Schultz AH, Starck D), pp. 1–252. Basel: Karger.

Seiler R (1977) Morphological and functional differentiation of muscles—studies on the m. frontalis, auricularis superior and auricularis anterior of primates including man. *Verh Anat Ges* 71, 1385–1388.

Seiler R (1979a) Criteria of the homology and phylogeny of facial muscles in primates including man, I, Prosimia and Platyrrhina. *Morphol Jahrb* 125, 191–217.

Seiler R (1979b) Criteria of the homology and phylogeny of facial muscles in primates including man, II, Catarrhina. *Morphol Jahrb* 125, 298–323.

Seiler R (1980) Ontogenesis of facial muscles in primates. *Morphol Jahrb* 126, 841–864.

Shoshani J, Groves CP, Simons EL, Gunnell GF (1996) Primate phylogeny: morphological vs molecular results. *Mol Phylogenet Evol* 5, 102–154.

Shrewsbury MM, Marzke MM, Linscheid RL, Reece SP (2003) Comparative morphology of the pollical distal phalanx. *Am J Phys Anthropol* 121, 30–47.

Shrivastava RK (1978) *Anatomie comparée du muscle deltoide et son innervation dans la série des mammifères*. Unpublished Phd thesis, Université de Paris, Paris.

Sigmon BA (1974) A functional analysis of pongid hip and thigh musculature. *J Hum Evol* 3, 161–185.

Smith WC (1923) The levator ani muscle; its structure in man, and its comparative relationships. *Anat Rec* 26, 175–204.

Sneath RS (1955) The insertion of the biceps femoris. *J Anat* 89, 550–553.

Sonntag CF (1923) On the anatomy, physiology, and pathology of the chimpanzee. *Proc Zool Soc Lond* 23, 323–429.

Sonntag CF (1924) *The morphology and evolution of the apes and man*. London: John Bale Sons and Danielsson, Ltd.

Sperino G (1897) *Anatomia del cimpanzè (Anthropopithecus Troglodytes)*. Torino: Unione Tipografica.

Starck D, Schneider R (1960) Respirationsorgane. In *Primatologia III/2* (eds. Hofer H, Schultz AH, Starck D), pp. 423–587. Basel: Karger.

Starck D (1973) The skull of the fetal chimpanzee. In *The Chimpanzee, Vol. 6*. (ed. Bourne GH), pp. 1–33. Basel: Karger.

Stern JT (1972) Anatomical and functional specializations of the human gluteus maximus. *Anta J Phys Anthropol* 36, 315–340.

Stern JT, Larson SG (2001) Telemetered electromyography of the supinators and pronators of the forearm in gibbons and chimpanzees: implications for the fundamental positional adaptation of hominoids. *Am J Phys Anthropol* 115, 253–268.

Stern JT, Wells JP, Jungers WL, Vangor AK, Fleagle JG (1980a) An electromyographic study of the pectoralis major in atelines and *Hylobates* with special reference to the evolution of a pars clavicularis. *Am J Phys Anthropol* 52, 13–25.

Stern JT, Wells JP, Jungers WL, Vangor AK (1980b) An electromyographic study of serratus anterior in atelines and *Alouatta*: implications for hominoid evolution. *Am J Phys Anthropol* 52, 323–334.

Stewart TD (1936) The musculature of the anthropoids, I, neck and trunk. *Am J Phys Anthropol* 21, 141–204.

Straus WL (1941a) The phylogeny of the human forearm extensors. *Hum Biol* 13, 23–50.
Straus WL (1941b) The phylogeny of the human forearm extensors (concluded). *Hum Biol* 13, 203–238.
Straus WL (1942a) The homologies of the forearm flexors: urodeles, lizards, mammals. *Am J Anat* 70, 281–316.
Straus WL (1942b) Rudimentary digits in primates. *Q Rev Biol* 17, 228–243.
Sullivan WE, Osgood CW (1925) The facialis musculature of the orang, *Simia satyrus*. *Anat Rec* 29, 195–343.
Susman RL (1994) Fossil evidence for early hominid tool use. *Science* 265, 1570–1573
Susman RL (1998) Hand function and tool behavior in early hominids. *J Hum Evol* 35, 23–46.
Susman RL, Stern JT (1980) EMG of the interosseous and lumbrical muscles in the chimpanzee (*Pan troglodytes*) hand during locomotion. *Am J Anat* 157, 389–397.
Susman RL, Jungers WL, Stern JT (1982) The functional morphology of the accessory interosseous muscle in the gibbon hand: determination of locomotor and manipulatory compromises. *J Anat* 134, 111–120.
Susman RL, Nyati L, Jassal MS (1999) Observations on the pollical palmar interosseus muscle (of Henle). *Anat Rec* 254, 159–165.
Sutton JB (1883) On some points in the anatomy of the chimpanzee (*Anthropopithecus troglodytes*). *J Anat Physiol* 18, 66–85.
Symington J (1889) Observations on the myology of the gorilla and chimpanzee. *Rep Brit Assoc Adv Sci* 59, 629–630.
Swindler DR, Wood CD (1973) *An atlas of primate gross anatomy: baboon, chimpanzee and men*. Seattle: University of Washington Press.
Tappen NC (1955) Relative weights of some functionally important muscles of the thigh, hip and leg in a gibbon and in man. *Am J Phys Anthropol* 13, 415–420.
Testut L (1883) Le long fléchisseur propre du pouce chez l'homme et les singes. *Bull Soc Zool Fr* 8, 164–185.
Testut L (1884) *Les anomalies musculaires chez l'homme expliquèes par l'anatomie comparée et leur importance en anthropologie*. Paris: Masson.
Thompson P (1901) On the arrangement of the fasciae of the pelvis and their relationship to the levator ani. *J Anat Phys* 35, 127–141.
Thorpe SK, Crompton RH, Günther MM, Ker RF, Alexander RM (1999) Dimensions and moment arms of the hind- and forelimb muscles of common chimpanzees (*Pan troglodytes*). *Am J Phys Anthropol* 110, 179–199.
Tocheri MW, Orr CM, Jacofsky MC, Marzke MW (2008) The evolutionary history of the hominin hand since the last common ancestor of *Pan* and *Homo*. *J Anat* 212, 544–562.
Tschachmachtschjan H (1912) Über die Pectoral- und Abdominal- musculatur und über die Scalenus-Gruppe bei Primataten. *Morph Jb* 44, 297–370.
Tuttle RH (1967) Knuckle-walking and the evolution of hominoid hands. *Am J Phys Anthrop* 26, 171–206.
Tuttle RH (1969) Quantitative and functional studies on the hands of the Anthropoidea, I, the Hominoidea. *J Morphol* 128, 309–363.
Tuttle RH (1970) Postural, propulsive and prehensile capabilities in the cheiridia of chimpanzees and other great apes. In *The Chimpanzee, Vol. 2* (ed. Bourne GH), pp. 167–263. Basel: Karger.
Tuttle RH (1972) Relative mass of cheiridial muscles in catarrhine primates. In *The Functional and Evolutionary Biology of Primates* (ed. Tuttle RH), pp. 262–291. Chicago: Aldine-Atherdon.
Tuttle RH, Basmajian JV (1974a) Electromyography of the brachial muscles in *Pan gorilla* and hominoid evolution. *Am J Phys Anthropol* 41, 71–90.
Tuttle RH, Basmajian JV (1974b) Electromyography of forearm musculature in *Gorilla* and problems related to knuckle-walking. In *Primate Locomotion* (ed. Jenkins FA), pp. 293–347. New York: Academic Press.

Tuttle RH, Basmajian JV (1976) Electromyography of pongid shoulder muscles and hominoid evolution I—retractors of the humerus and rotators of the scapula. *Yearbook Phys Anthropol* 20, 491–497.

Tuttle RH, Basmajian JV (1978a) Electromyography of pongid shoulder muscles II—deltoid, rhomboid and «rotator cuff». *Am J Phys Anthropol* 49, 47–56.

Tuttle RH, Basmajian JV (1978b) Electromyography of pongid shoulder muscles III—quadrupedal positional behavior. *Am J Phys Anthropol* 49, 57–70.

Tuttle RH, Velte MJ, Basmajian JV (1983) Electromyography of brachial muscles in *Pan troglodytes* and *Pongo pygmaeus*. *Am J Phys Anthropol* 61, 75–83.

Tuttle RH, Hollowed JR, Basmajian JV (1992) Electromyography of pronators and supinators in great apes. *Am J Phys Anthropol* 87, 215–26.

Tyson E (1699) *Orang-Outang sive Homo sylvestris, or the anatomy of a pygmie compared to that of a monkey, an ape and a man*. London: T. Bennet.

Vallois H (1914) *Étude anatomique de l'articulation du genou ches les Primates*. Montpellier: L'Abeille.

Van den Broek AJP (1908) Ueber einige anatomische Merkmale von *Ateles* in Zusammenhang mit der Anatomie der Platyrrhinin. *Anat Anz* 33, 111–124.

Van den Broek AJP (1914) Studien zur Morphologie des Primatenbeckens. *Morphol Jahrb* 49, 1–118.

Van Westrienen A (1907) Das Kniegelenk der Primaten, mit besonderer Berücksichtigung der Anthropoiden. *Petrus Camper* 4, 1–60.

Walmsley R (1937) The sheath of the rectus abdominis. *J Anat Phys* 77, 404–414.

Verhulst J (2003) *Developmental dynamics in humans and other primates: discovering evolutionary principles through comparative morphology*. Ghent: Adonis Press.

Virchow H (1915) Gesichtsmuskeln des Schimpansen. *Abh Preuss Akad Wiss* 1, 1–81.

Vrolik W (1841) *Recherches d' anatornie comparé, sur le chimpanzé*. Amsterdam: Johannes Miller.

Waller BM, Vick SJ, Parr LA, Bard KA, Smith Pasqualini MC, Gothard K, Fuglevand A (2006) Intramuscular electrical stimulation of facial muscles in humans and chimpanzees: Duchenne revisited and extended. *Emotion* 6, 367–382.

Waller BM, Burrows AM, Cray JJ (2008a) Selection for universal facial emotion. *Emotion* 8, 435–439.

Warnots L (1885) Identité de la musculature du larynx chez l'homme et chez le chimpansé. *Bull soc Anthropol Bruxelles* 4, 59–62.

Waterman HC (1929) Studies on the evolution of the pelvis of man and other primates. *Bull Am Mus Nat Hist* 55, 585–642.

Whitehead PF (1993) Aspects of the anthropoid wrist and hand. In *Postcranial Adaptation in Nonhuman Primates* (ed. Gebo DL), pp 96–120. DeKalb: Northern Illinois University Press.

Wilder B (1862) Contributions to the comparative myology of the chimpanzee. *Boston J Nat Hist* 6, 352–384.

Winckler G (1947) Recherches sur le systéme musculaire du transverso-spinalis. *Arch Anat Histol Embryol* 30, 151–231.

Winckler G (1950) Contribution a l'étude des muscles larges de la paroi abdominale, Étude d'anatomie comparée. *Arch Anat Histol Embryol* 33, 157–228.

Windle BCA (1886a) Notes on the myology of *Midas rosalia*, with remarks on the muscular system of apes. *Proc Bgham Nat Hist Soc* 5, 152–166.

Wood Jones F (1920) *The principles of anatomy as seen in the hand*. London: J. & A Churchill.

Wyman J (1855) An account of the dissection of a black chimpanzee. *Proc Boston Soc Nat Hist* 5, 270–275.

Yirga S (1987) Interrelation between ischium, thigh extending muscles and locomotion in some primates. *Primates* 28, 79–86.

Ziegler AC (1964) Brachiating adaptations of chimpanzee upper limb musculature. *Am J Phys Anthropol* 22, 15–32.

Zihlman AL, Brunker L (1979) Hominid bipedalism: then and now. *Yearb Phys Anthropol* 22, 132–162.

Appendix II
Literature Cited, Not Including Information about the Muscles of Chimpanzees

Diogo R (2004a) *Morphological evolution, aptations, homoplasies, constraints, and evolutionary trends: catfishes as a case study on general phylogeny and macroevolution.* Enfield: Science Publishers.
Diogo R (2004b) Muscles versus bones: catfishes as a case study for an analysis on the contribution of myological and osteological structures in phylogenetic reconstructions. *Anim Biol* 54, 373–391.
Diogo R (2007) *On the origin and evolution of higher-clades: osteology, myology, phylogeny and macroevolution of bony fishes and the rise of tetrapods.* Enfield: Science Publishers.
Diogo R (2008) Comparative anatomy, homologies and evolution of the mandibular, hyoid and hypobranchial muscles of bony fish and tetrapods: a new insight. *Anim Biol* 58, 123–172.
Diogo R (2009) The head musculature of the Philippine colugo (Dermoptera: *Cynocephalus volans*), with a comparison to tree-shrews, primates and other mammals. *J Morphol* 270, 14–51.
Diogo R, Abdala V (2007) Comparative anatomy, homologies and evolution of the pectoral muscles of bony fish and tetrapods: a new insight. *J Morphol* 268, 504–517.
Diogo R, Potau JM, Pastor JF, de Paz FJ, Ferrero EM, Bello G, Barbosa M, Wood B (2010) Photographic and Descriptive Musculoskeletal Atlas of *Gorilla*—with notes on the attachments, variations, innervation, synonymy and weight of the muscles. Oxford: Taylor & Francis.
Diogo R, Potau JM, Pastor JF, de Paz FJ, Ferrero EM, Bello G, Barbosa M, Aziz MA, Burrows AM, Arias-Martorell J, Wood B (2012) Photographic and Descriptive Musculoskeletal Atlas of Gibbons and Siamangs (*Hylobates*)—with notes on the attachments, variations, innervation, synonymy and weight of the muscles. Oxford: Taylor & Francis.
Netter FH (2006) *Atlas of human anatomy (4th ed.).* Philadelphia: Saunders.
Terminologia Anatomica (1998) *Federative Committee on Anatomical Terminology.* Stuttgart: Georg Thieme. New York: Columbia University Press.
Wood J (1870) On a group of varieties of the muscles of the human neck, shoulder, and chest, with their transitional forms and homologies in the Mammalia. *Philos Trans R Soc Lond* 160, 83–116.

Index

A

Abductor digiti minimi 56, 88
Abductor hallucis 85, 88, 89, 91
Abductor os metatarsi digiti quinti 89
Abductor ossis metacarpi hallucis 89
Abductor pollicis brevis 54–56
Abductor pollicis longus 61
Adductor brevis 81, 82
Adductor hallucis 90, 91
Adductor longus 81, 82
Adductor magnus 79, 81, 82
Adductor minimus 82
Adductor pollicis 50–54, 62
Adductor pollicis accessorius 50, 51
Anconeus 41, 47, 58
Anomalus maxillae 18
Anomalus menti 18
Anomalus nasi 18
Antitragicus 9
Articularis cubiti 41
Articularis genu 80
Arytenoideus obliquus 26
Arytenoideus transverses 26
Atlantomastoideus 67
Atlantoscapularis posticus 33
Auricularis anterior 12, 13
Auricularis posterior 8, 9
Auricularis superior 13
Auriculo-orbitalis 11–13

B

Biceps brachii 34, 41–43
Biceps femoris 77, 82–84
Brachialis 41, 42
Brachioradialis 57

Buccinatorius 15, 18
Bulbospongiosus 74

C

Caput cleido-occipitale 20
Ceratoarytenoideus lateralis 25
Ceratocricoideus 26
Ceratoglossus 28
Ceratohyoideus 19
Chondroglossus 28
Cleido-occipitalis 20
Coccygeus 73
Compressor urethrae 75
Constrictor pharyngis inferior 21
Constrictor pharyngis medius 21
Constrictor pharyngis superior 22
Contrahentes digitorum 49
Coracobrachialis 40, 42, 43
Coracobrachialis medius/coracobrachialis
    proprius 43
Coracobrachialis profundus/coracobrachialis
    brevis 43
Coracobrachialis superficialis/longus 43
Corrugator supercilii 14
Costocoracoideus 34
Cremaster 71
Cricoarytenoideus lateralis 26
Cricoarytenoideus posterior 2, 25–27
Cricothyroideus 22, 30
Cuspidator oris 18

D

Deep additional slip of the adductor pollicis 50
Deep head of the flexor pollicis brevis 51–54
Deltoideus 36–38
Depressor anguli oris 18, 19

Depressor glabellae 15
Depressor helicis 9
Depressor labii inferioris 18, 19
Depressor septi nasi 16, 17
Depressor supercilii 14
Depressor tarsi 11
Detrusor vesicae 75
Diaphragma 69
Digastricus anterior 4
Digastricus posterior 7
Dorsoepitrochlearis 40, 43

E

Epitrochleoanconeus 47
Erector spinae 66, 67
Extensor brevis digitorum manus 60
Extensor carpi radialis brevis 56
Extensor carpi radialis longus 56
Extensor carpi ulnaris 57, 58
Extensor communis pollicis et indicis 60
Extensor digiti III proprius 60
Extensor digiti minimi 58
Extensor digiti quarti 59
Extensor digitorum 58–61, 84, 85, 88
Extensor digitorum brevis 88
Extensor digitorum longus 58, 84
Extensor hallucis brevis 88
Extensor hallucis longus 84
Extensor indicis 59, 60
Extensor pollicis brevis 61, 62
Extensor pollicis longus 60

F

Fibularis brevis 85, 88
Fibularis longus 85, 88, 90, 91
Fibularis tertius 85
Flexor brevis profundus 1 51–53
Flexor brevis profundus 2 50–54, 56
Flexor brevis profundus 10 52
Flexor carpi radialis 47
Flexor carpi ulnaris 46, 47
Flexor caudae 73
Flexor digiti mimini brevis 90
Flexor digitorum brevis 49, 87, 89
Flexor digitorum brevis manus 49
Flexor digitorum longus 87, 89
Flexor digitorum profundus 44, 45, 49
Flexor digitorum superficialis 45, 46
Flexor hallucis brevis 91

Flexor hallucis longus 87, 89, 90
Flexor pollicis brevis 51–54, 56
Flexor pollicis longus 44, 45
Flexores breves profundi 49, 51, 52
Frontalis 8, 11–13, 15

G

Gastrocnemius 86
Gemellus inferior 78
Gemellus superior 78
Genio-epiglotticus 27
Genioglossus 27
Genio-hyo-epiglotticus 27
Genio-hyoglossus 27
Geniohyoideus 3, 27
Glosso-epiglotticus 27
Gluteus maximus 73, 76–78, 83
Gluteus medius 76, 77, 79
Gluteus minimus 77, 78
Gracilis 81–83

H

Helicis major 9
Helicis minor 9
Hyo-epiglotticus 27
Hyoglossus 27, 28

I

Iliacus 72, 76
Iliocapsularis 83
Iliococcygeus 73
Iliocostalis 66, 67
Incisivus labii inferioris 18
Incisivus labii superior 18
Incisivus labii superioris 18
Incisurae Santorini 9
Incisurae terminalis 9
Infraorbitalis 14
Infraspinatus 36–38
Intercapitulares 49
Intercartilagineus 9
Intercostales externi 65
Intercostales interni 65
Interdigitales 53
Intermandibularis anterior 4
Intermetacarpales 51, 52, 54
Interossei accessorii 52
Interossei dorsales 52, 91, 92
Interossei palmares 52, 54

Interossei plantares 91, 92
Interosseous volaris primus of Henle 50
Interspinales 68
Intertransversarii 68
Intrinsic facial muscles of ear 9
Intrinsic muscles of tongue 27
Ischiocavernosus 74
Ischiofemoralis 78

J

Jugulohyoideus 6

L

Labialis inferior profundus 18
Labialis superior profundus 16–18
Latissimus dorsi 39, 40, 71
Levator anguli oris facialis 10, 11, 17, 19
Levator anguli oris mandibularis 17
Levator ani 73, 74
Levator claviculae 33
Levator glandulae thyroideae 29, 30
Levator labii superioris 14, 15
Levator labii superioris alaeque nasi 14, 15
Levator palpebrae superioris 30
Levator scapulae 31–33, 65
Levator veli palatine 5, 23
Levatores costarum 65
Longissimus 66–68
Longitudinalis inferior 27
Longitudinalis superior 27
Longus capitis 63, 64
Longus colli 64
Lumbricales 49, 51, 52, 89, 90

M

Mandibulo-auricularis 9
Masseter 5
Mentalis 19
Multifidus 68
Muscles of eye 30
Musculus nasalis impar 16
Musculus uvulae 23
Musculus ventricularis laryngis 24
Musculus vocalis 24–26
Mylohyoideus 4

N

Nasalis 16, 17

O

Obliquus auriculae 9
Obliquus capitis inferior 63
Obliquus capitis superior 63
Obliquus externus abdominis 71
Obliquus inferior 30
Obliquus internus abdominis 71
Obliquus superior 30
Obturatorius externus 78
Obturatorius internus 79
Occipitalis 8, 20, 32
Omohyoideus 28, 29
Opponens digiti minimi 52, 55, 90, 91
Opponens hallucis 91
Opponens pollicis 51–54
Orbicularis oculi 10, 11, 13, 14
Orbicularis oris 10, 16–19
Orbitalis 11–14, 30

P

Palatoglossus 28
Palatopharyngeal sphincter 22
Palatopharyngeus 22, 23
Palmaris brevis 48, 49
Palmaris longus 46
Palmaris superficialis 49
Pars aryepiglottica 25
Pars arymembranosa 25
Pars supralabialis of the buccinatorius 18
Pars thyroepiglottica 25
Pars thyromembranosa 25
Passavant's ridge 22
Pectineus 83
Pectoralis major 34, 35, 71
Pectoralis tertius 36
Petropharyngeus 19
Piriformis 77, 79
Plantaris 86
Platysma myoides 7–9, 19
Popliteus 87
Preorbicularis 11
Procerus 15
Pronator quadratus 43, 44
Pronator teres 42, 48
Psoas major 70, 76
Psoas minor 76
Pterygoideus lateralis 6
Pterygoideus medialis 6
Pterygopharyngeus 22

Pterygotympanicus 5
Pubococcygeus 73, 74
Puboprostaticus 75
Puborectalis 74
Pubovesicalis 73
Pyramidalis 9, 15, 70
Pyramidalis auriculae 9

Q

Quadratus femoris 78, 79
Quadratus lumborum 72
Quadratus plantae 89

R

Rectococcygeus 74
Rectourethralis 74
Rectovesicalis 75
Rectus abdominis 70
Rectus capitis anterior 63
Rectus capitis lateralis 63
Rectus capitis posterior major 63
Rectus capitis posterior minor 63, 64
Rectus femoris 80, 81
Rectus inferior 30
Rectus labii inferioris 18
Rectus labii superioris 18
Rectus lateralis 30
Rectus medialis 30
Rectus superior 30
Regionis analis 75
Regionis urogenitalis 75
Rhomboideus 31, 32
Rhomboideus major 32
Rhomboideus minor 32
Rhomboideus occipitalis 32
Risorius 8–10
Rotatores 68

S

Salpingopharyngeus 23
Sartorius 71, 81
Scalenus anterior 63–65
Scalenus medius 64, 65
Scalenus minimus 65
Scalenus posterior 64, 65
Scansorius 77, 78
Scapuloclavicularis 37
Semimembranosus 83, 84
Semispinalis capitis 68
Semispinalis cervicis 67

Semispinalis thoracis 67
Semitendinosus 81–84
Serratus anterior 31, 71
Serratus posterior inferior 66
Serratus posterior superior 66
Soleus 86
Sphincter ampullae 75
Sphincter ani externus 74
Sphincter ani internus 75
Sphincter colli profundus 7, 8
Sphincter colli superficialis 7, 8
Sphincter ductus choledochi 75
Sphincter palatopharyngeus 22
Sphincter pyloricus 75
Sphincter urethrae 74
Sphincter urethrovaginalis 75
Spinalis 38, 66–68
Splenius capitis 65
Splenius cervicis 66
Stapedius 7
Sternocleidomastoideus 20, 21, 29
Sternohyoideus 2, 28
Sternothyroideus 29
Styloglossus 28
Stylohyoideus 1, 6
Stylolaryngeus 7
Stylopharyngeus 19
Subclavius 33, 34
Subnasalis 16, 17
Subscapularis 36, 38
Superficial head of the flexor pollicis brevis
    51–54
Supinator 57
Supracostalis 70
Supraspinatus 36, 37
Suspensori duodeni 75

T

TDAS-AD 50
Temporalis 5, 6, 10, 12, 13
Temporoparietalis 11–13
Tensor fasciae latae 81
Tensor linea semilunaris 70
Tensor tympani 4
Tensor veli palatine 5, 23
Teres major 38, 39
Teres minor 37, 38
Thin 13, 21, 44, 50
Thryroarytenoideus lateralis 24
Thyroarytenoideus 24–26
Thyroarytenoideus inferior 25

Thyroarytenoideus medialis 25
Thyroarytenoideus superior 25
Thyrohyoideus 28, 29
Thyroideus transverses 22
Tibialis anterior 85
Tibialis posterior 87, 91
Tragicus 9
Trago-helicinus 9
Transversalis pedis 90
Transversus abdominis 70, 71
Transversus auriculae 9
Transversus linguae 27
Transversus menti 19
Transversus perinei profundus 74
Transversus perinei superficialis 74
Transversus thoracis 65
Trapezius 20, 33
Triceps brachii 40, 41
Trigoni vesicae 75

U

Ulnaris-quinti 58

V

Vastus intermedius 80, 81
Vastus lateralis 77, 78, 80
Vastus medialis 80
Ventricularis 24, 25
Verticalis linguae 27
Vesicoprostaticus 75
Vesicovaginalis 75

X

Xiphihumeralis 36

Z

Zygomatico-auricularis 13
Zygomatico-mandibularis 5
Zygomatico-orbicularis 14
Zygomaticus major 10, 11
Zygomaticus minor 10, 11

About The Authors

Rui Diogo is an Assistant Professor at the Howard University College of Medicine and a Resource Faculty at the Center for the Advanced Study of Hominid Paleobiology of George Washington University (US). He is the author or co-author of numerous publications, and the co-editor of the books *Catfishes* and *Gonorynchiformes and ostariophysan interrelationships—a comprehensive review*. He is the sole author or first author of the books *Morphological evolution, aptations, homoplasies, constraints and evolutionary trends—catfishes as a case study on general phylogeny and macroevolution, The origin of higher clades—osteology, myology, phylogeny and evolution of bony fishes and the rise of tetrapods Muscles of vertebrates—comparative anatomy, evolution, homologies and development, Photographic and descriptive musculoskeletal atlas of Gorilla—with notes on the attachments, variations, innervation, synonymy and weight of the muscles, Photographic and descriptive musculoskeletal atlas of gibbons and siamangs* (Hylobates)—*with notes on the attachments, variations, innervation, synonymy and weight of the muscles* and *Comparative anatomy and phylogeny of primate muscles and human evolution*.

Josep Potau is Professor at the Department of Anatomy and Embryology of the University of Barcelona (Spain) and is the director of the University's Center for the Study of Comparative and Evolutionary Anatomy. His current research focus on the analysis of functional and anatomical adaptations associated with the evolution of different types of locomotion and of the upper limb musculature within primates. He has published several papers and book chapters on functional and comparative anatomy.

Juan Pastor is Professor at the Department of Anatomy of the University of Valladolid (Spain) and is the director of the University's Anatomical Museum, which houses the largest comparative osteological collection in Spain. He published several papers on comparative anatomy and anthropology.

Félix de Paz is Professor at the Department of Anatomy of the University of Valladolid and is a member of the Royal Academy of Medicine and Surgery of Valladolid (Spain). He has published several papers on comparative anatomy and anthropology.

Mercedes Barbosa is Professor at the Department of Anatomy of the University of Valladolid (Spain), and is a member of the Anatomical Society of Spain. She published several papers on physical anthropology.

Eva Ferrero is a biologist who graduated from the University of León (Spain) and obtained her PhD at the University of Valladolid (Spain) where she is undertaking research on the comparative anatomy of primates and other mammals.

Gaëlle Bello is a biologist who graduated from the University of La Coruña (Spain). She undertook a Master in Primatology at the University of Barcelona (Spain), and is now undertaking a PhD at the University of Barcelona that focuses on the evolution of the scapula within Primates and its adaptations to different types of locomotion.

Anne M. Burrows is a biological anthropologist and an Associate Professor in the Department of Physical Therapy at Duquesne University and a Research Associate Professor in the Department of Anthropology at the University of Pittsburgh. Her current research focuses on the evolution of facial musculature in primates and the evolution of feeding specializations in primates. She has published numerous papers and book chapters on these subjects and has recently co-edited *The Evolution of Exudativory in Primates*.

M. Ashraf Aziz is Professor of Anatomy at the Department of Anatomy of Howard University College of Medicine (USA). His research focuses on the comparative gross and developmental morphology of modern human aneuploidy syndromes, the evolution of the muscles supplied by the trigeminal nerve and of the arm and hand muscles in non-human and human primates and the value of human cadaver dissections/prosections in the age of digital information systems. He has published numerous papers in international journals, including *Teratology, American Journal of Physical Anthropology, Journal of Anatomy, The Anatomical Record* and *Primates*.

Julia Arias-Martorell is a biologist now undertaking a PhD at the University of Barcelona (Spain) that focuses on functional morphology and variabilty of the forelimbs of the hominoids related to the diverse locomotor repertoires of the members of this clade, and on the enhancement of tridimensional techniques to study evolution.

Bernard Wood is University Professor of Human Origins and Director of the Center for the Advanced Study of Hominid Paleobiology at George Washington University (USA). His edited publications include *Food Acquisition and Processing in Primates* and *Major Topics in Primate and Human Evolution* and he is the author of *The Evolution of Early Man, Human Evolution, Koobi Fora Research Project—Hominid Cranial Remains (Vol. 4), Human Evolution—A Very Short Introduction*. He is the editor of the *Wiley-Blackwell Encyclopedia of Human Evolution*.

Color Plate Section

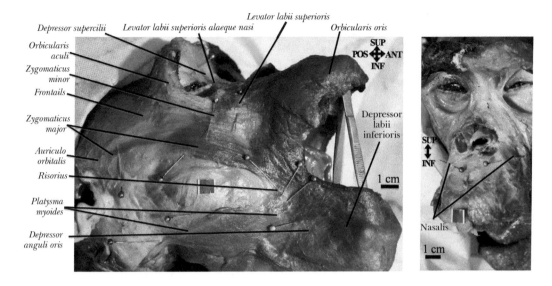

Fig. 1 *Pan troglodytes* (VU PT1, adult male). On the left: lateral view of the left facial musculature showing the risorius, which, interestingly, is not differentiated into a separate muscle on the right side of the specimen. On the right: frontal view of the deep facial musculature, showing the nasalis on both sides of the specimen. In this figure and the remaining figures of the atlas, the names of the muscles are in italics, and SUP, INF, ANT, POS, MED, LAT, VEN, DOR, PRO and DIS refer to superior, inferior, anterior, posterior, medial, lateral, ventral, dorsal, proximal and distal, respectively (in the sense the terms are applied to pronograde tetrapods: see Methodology and Material).

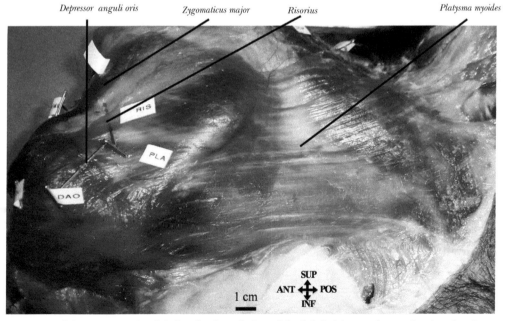

Fig. 2 *Pan troglodytes* (PFA 1016, adult female): lateral view of the left facial musculature, showing the platysma myoides and a thin muscular structure that seems to correspond to the risorius of humans.

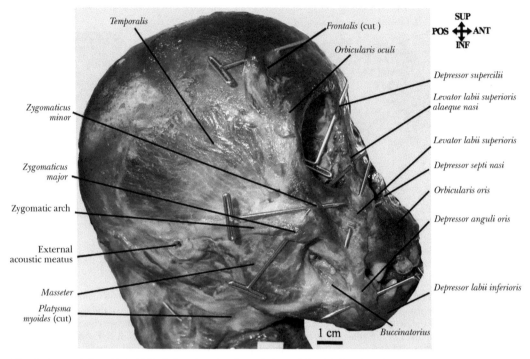

Fig. 3 *Pan troglodytes* (PFA 1051, infant female): lateral view of the left head and neck musclature after cutting some facial muscles.

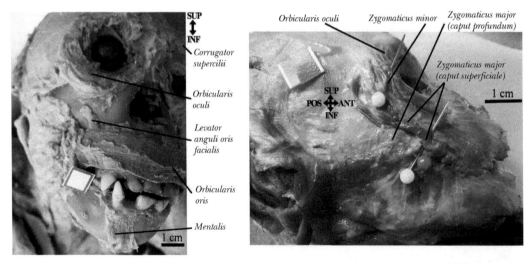

Fig. 4 On the left: *Pan troglodytes* (HU PT1, infant male), frontal view of the deep facial muscles. On the right: *Pan troglodytes* (PFA 1077, infant female), lateral view of the right zygomaticus major, showing the two heads of this muscle.

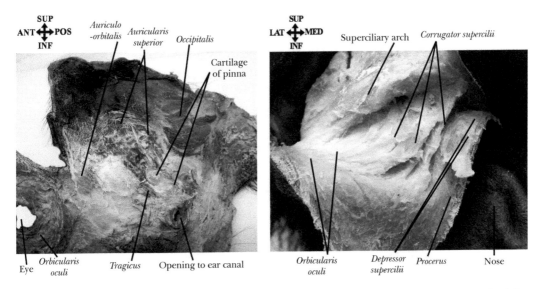

Fig. 5 *Pan troglodytes* (Yerkes uncatalogued, adult male). On the left: medial view of right facial mask. On the right: anterior view of the facial muscles of the eye region.

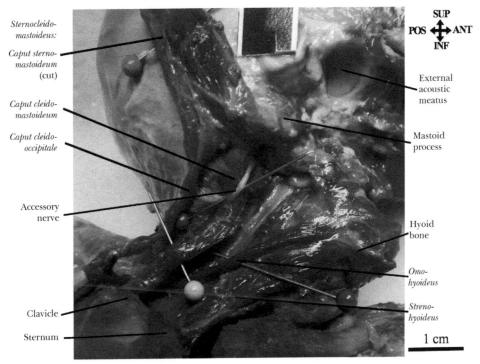

Fig. 6 *Pan troglodytes* (PFA 1077, infant female): lateral view of the left neck musculature after reflecting the caput sternomastoideum of the sternocleidomastoideus, which passes superficially to the accessory nerve, as does the caput cleido-occipitale, but not the caput cleidomastoideum.

120 *Photographic and Descriptive Musculoskeletal Atlas of Chimpanzees*

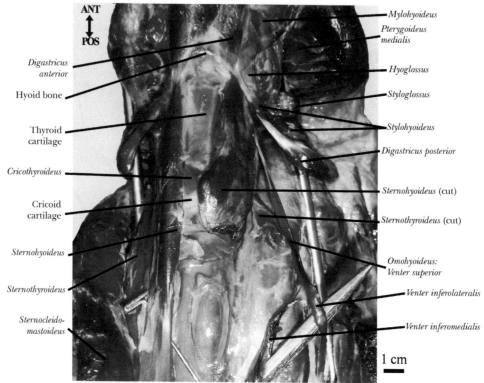

Fig. 7 *Pan troglodytes* (PFA 1016, adult female): ventral view of the head and neck musculature, showing the three heads of the omohyoideus, and the stylohyoideus pierced by the intermediate tendon of the digastric.

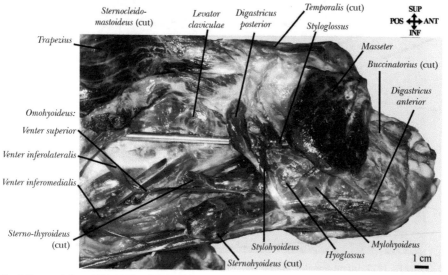

Fig. 8 *Pan troglodytes* (PFA 1016, adult female): ventrolateral view of the right head and neck musculature, showing the three heads of the omohyoideus, and the stylohyoideus pierced by the intermediate tendon of the digastric.

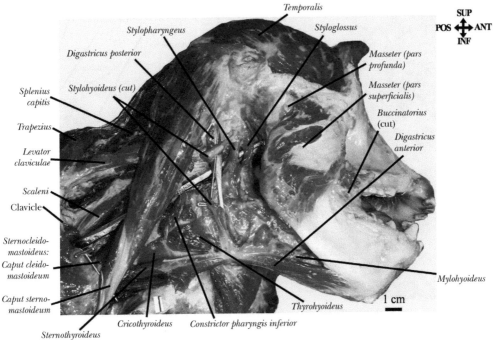

Fig. 9 *Pan troglodytes* (PFA 1009, adult female): ventrolateral view of the right head and neck musculature after cutting the stylohyoideus.

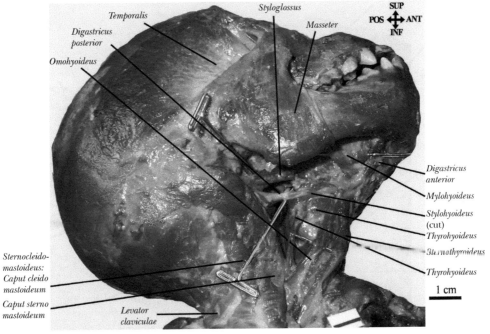

Fig. 10 *Pan troglodytes* (PFA 1051, infant female): ventrolateral view of the left head and neck musculature.

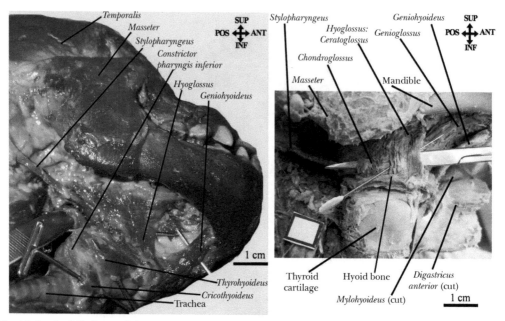

Fig. 11 On the left: *Pan troglodytes* (PFA 1051, infant female), ventrolateral view of the deep left head and neck musculature. On the right: *Pan troglodytes* (HU PT1, infant male): lateral view of the right hyoglossus (showing the division of this muscle into a ceratoglossus and a chondroglossus) after reflecting the mylohyoideus, geniohyoideus and digastricus anterior.

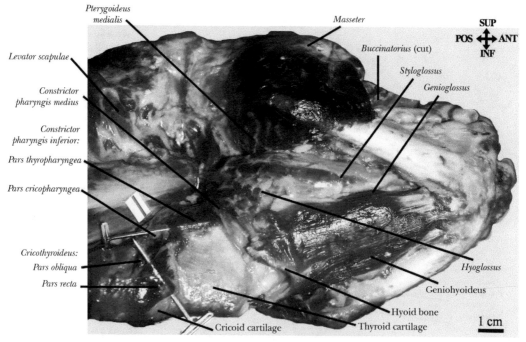

Fig. 12 *Pan troglodytes* (PFA 1016, adult female): ventrolateral view of the head and neck musculature showing the middle and inferior constrictors of the pharynx.

Color Plates 123

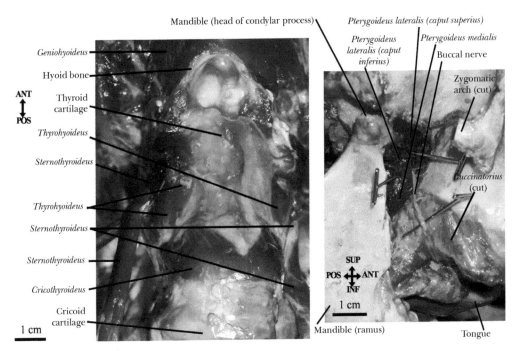

Fig. 13 *Pan troglodytes* (PFA 1016, adult female). On the left: ventral view of the head and neck musculature showing the anterior portion of the sternothyroideus extending anteriorly to the posterior portion of the thyrohyoideus. On the right: lateral view of the right buccal nerve mainly passing mainly between the caput superior and the caput inferius of the pterygoideus lateralis.

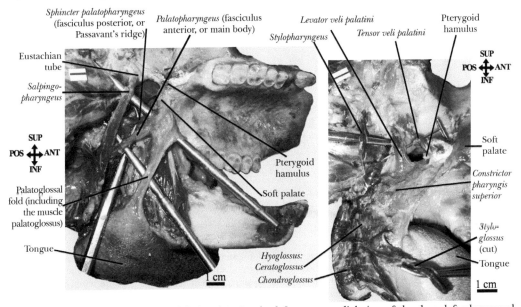

Fig. 14 *Pan troglodytes* (PFA 1016, adult female). On the left: ventromedial view of the deep left pharyngeal muscles. On the right: lateral view of the right tongue and pharyngeal muscles.

124 Photographic and Descriptive Musculoskeletal Atlas of Chimpanzees

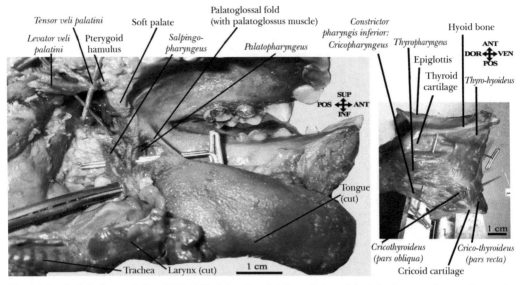

Fig. 15 On the left: *Pan troglodytes* (PFA 1051, infant female), lateral view of the left pharyngeal musculature. On the right: *Pan troglodytes* (PFA 1009, adult female), lateral view of the superficial musculature associated with the right side of the larynx.

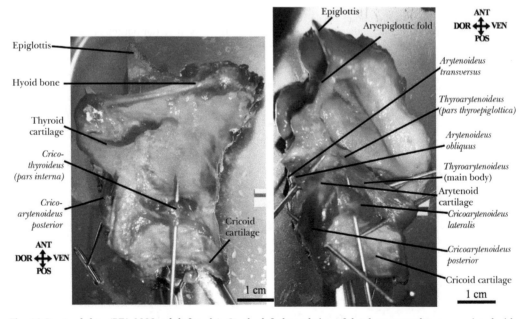

Fig. 16 *Pan troglodytes* (PFA 1009, adult female). On the left: lateral view of the deep musculature associated with the right side of the larynx. On the right: lateral view of the deep musculature associated with the right side of the larynx, the deep part of the cricothyroideus, as well as part of the thyroid cartilage, were removed in order to show the arytenoideus transversus, arytenoideus obliquus, cricoarytenoideus lateralis, and the main body and the pars thyroepiglottica of the thyroarytenoideus.

Color Plates 125

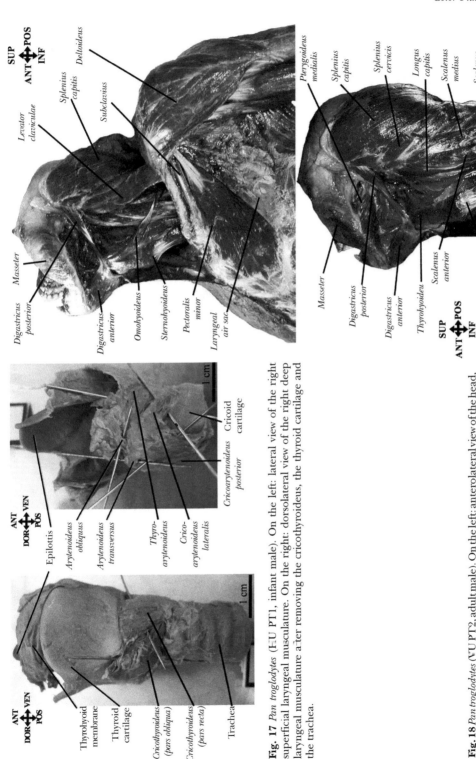

Fig. 17 *Pan troglodytes* (HU PT1, infant male). On the left: lateral view of the right superficial laryngeal musculature. On the right: dorsolateral view of the right deep laryngeal musculature after removing the cricothyroideus, the thyroid cartilage and the trachea.

Fig. 18 *Pan troglodytes* (VU PT2, adult male). On the left: anterolateral view of the head, neck, shoulder and arm muscles after removal of platysma, sternocleidomastoideus and pectoralis major. On the right: anterolateral view of the head and neck muscles after removal of the upper limb.

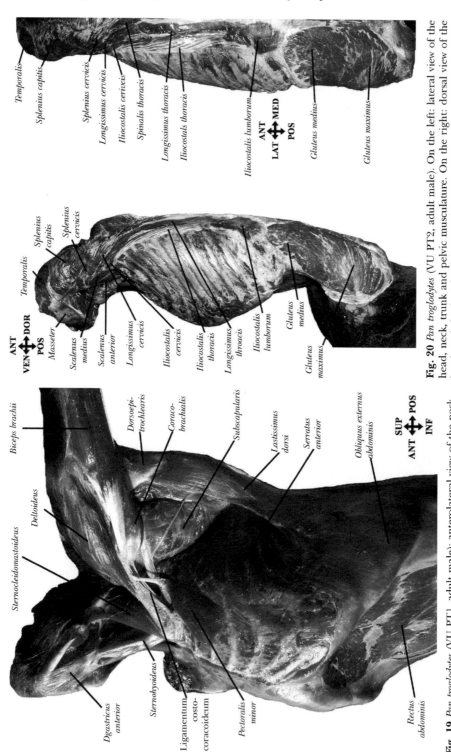

Fig. 19 *Pan troglodytes* (VU PT1, adult male): anterolateral view of the neck, thorax and shoulder muscles after removal of pectoralis major.

Fig. 20 *Pan troglodytes* (VU PT2, adult male). On the left: lateral view of the head, neck, trunk and pelvic musculature. On the right: dorsal view of the head, neck, back and pelvic musculature.

Color Plates 127

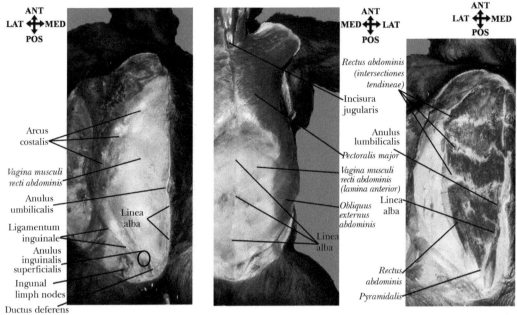

Fig. 21 *Pan troglodytes* (VU PT1, adult male). On the left: ventral view of the abdominal wall, after removing the skin. On the center: ventral view of the pectoral and abdominal musculature. On the right: ventral view of the right abdominal musculature after removing the anterior layer of the rectus sheath.

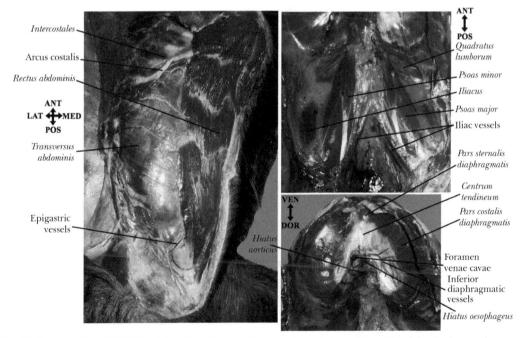

Fig. 22 *Pan troglodytes* (VU PT1, adult male). On the left: ventrolateral view of the right abdominal musculature after removing the anterior layer of the rectus sheath and the external and internal oblique muscles. On the right, top: ventral view of the posterior abdominal wall. On the right, bottom: posterior view of the diaphragma.

128 Photographic and Descriptive Musculoskeletal Atlas of Chimpanzees

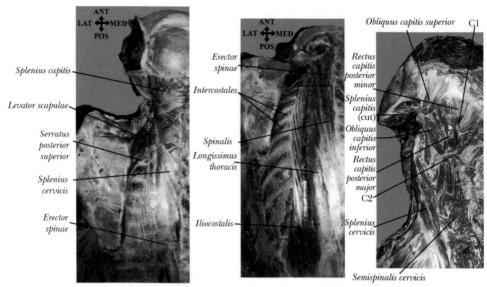

Fig. 23 On the left: *Pan troglodytes* (VU PT2, adult male), dorsal view of the back muscles, after removal of trapezius and rhomboids. On the center: *Pan troglodytes* (VU PT2, adult male), dorsal view of the deep back muscles, after removal of trapezius and latissimus dorsi. On the right: *Pan troglodytes* (VU PT1, adult male), dorsal view of the musculature of the posterior region of the neck.

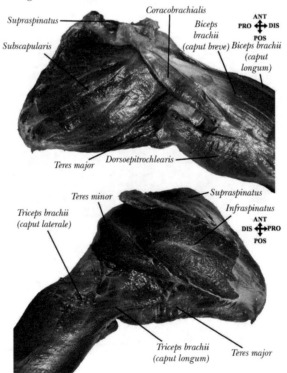

Fig. 24 *Pan troglodytes* (VU PT1, adult male). On the top: ventromedial view of the left shoulder and arm muscles. On the bottom: dorsolateral view of the left shoulder and arm muscles.

Color Plates 129

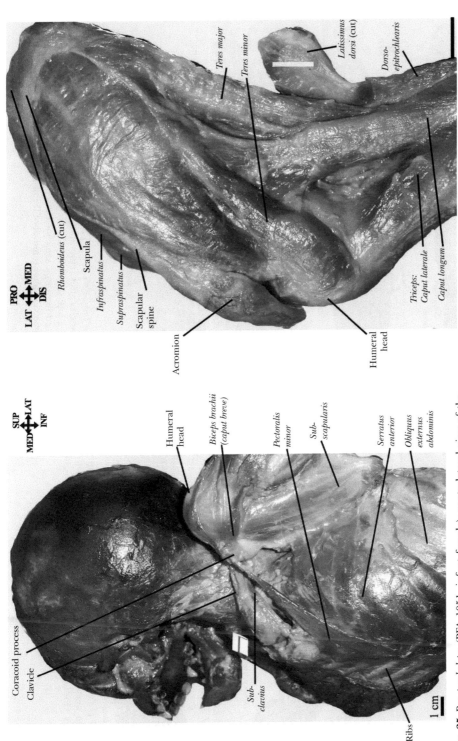

Fig. 25 *Pan troglodytes* (PFA 1051, infant female): ventrolateral view of the left pectoral and arm musculature after cutting the pectoralis major and deltoideus.

Fig. 26 *Pan troglodytes* (PFA 1051, infant female): dorsal view of the left pectoral and arm musculature after removal of deltoideus.

130 Photographic and Descriptive Musculoskeletal Atlas of Chimpanzees

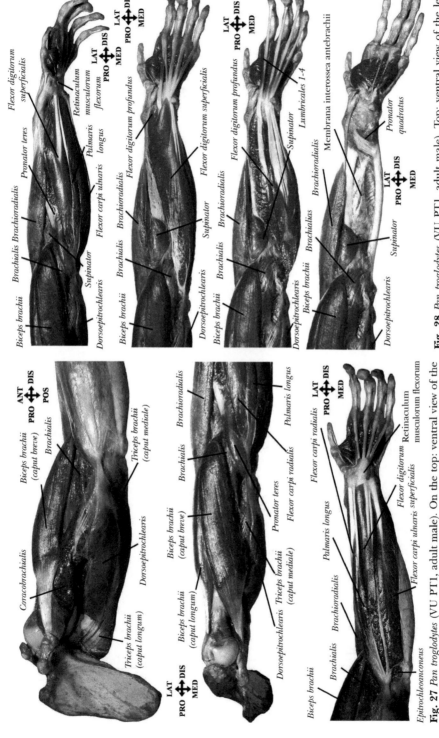

Fig. 28 *Pan troglodytes* (VU PT1, adult male). Top: ventral view of the left forearm muscles after removal of flexor carpi radialis and palmaris longus. Second from top: same view after removal of pronator teres and flexor carpi ulnaris. Third from top: same view after removal of digitorum superficialis. Bottom: same view after removal of flexor digitorum profundus.

Fig. 27 *Pan troglodytes* (VU PT1, adult male). On the top: ventral view of the left arm muscles. On the center: ventral view of the left arm and forearm muscles. On the bottom: ventral view of the left forearm muscles.

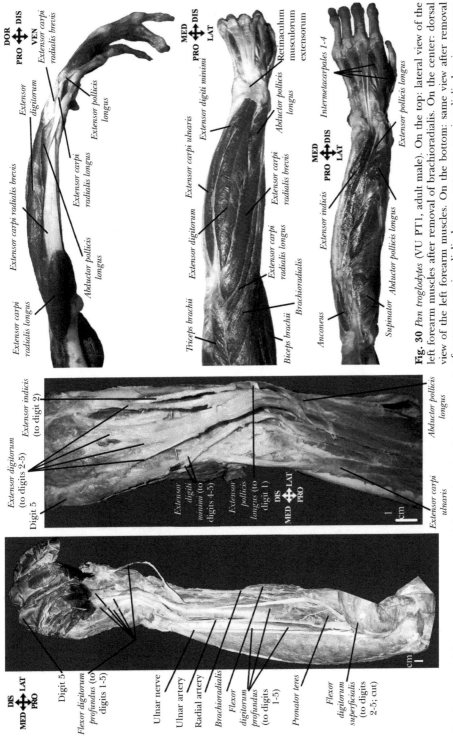

Fig. 29 *Pan troglodytes* (GWU-ANT PT2, adult female). On the left: ventral view of the deep right forearm muscles. On the right: dorsal view of the distal portion of the right forearm muscles.

Fig. 30 *Pan troglodytes* (VU PT1, adult male). On the top: lateral view of the left forearm muscles after removal of brachioradialis. On the center: dorsal view of the left forearm muscles. On the bottom: same view after removal of extensor carpi radialis longus, extensor carpi radialis brevis, extensor digitorum, extensor digiti minimi and extensor carpi ulnaris.

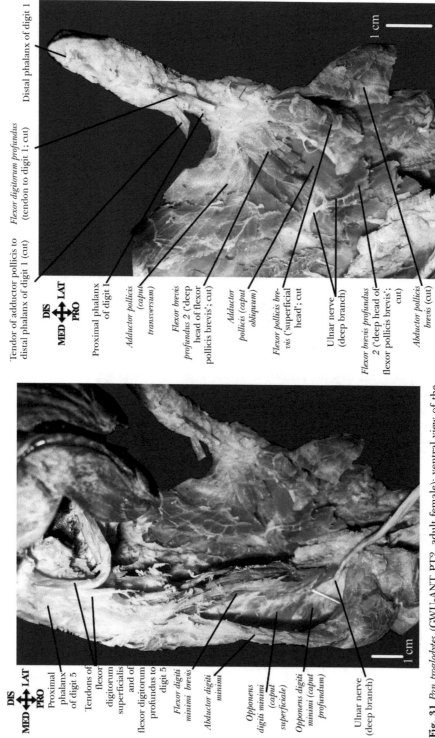

Fig. 31 *Pan troglodytes* (GWU-ANT PT2, adult female): ventral view of the right hypothenar muscles.

Fig. 32 *Pan troglodytes* (GWU-ANT PT2, adult female): ventral view of the right hypothenar muscles.

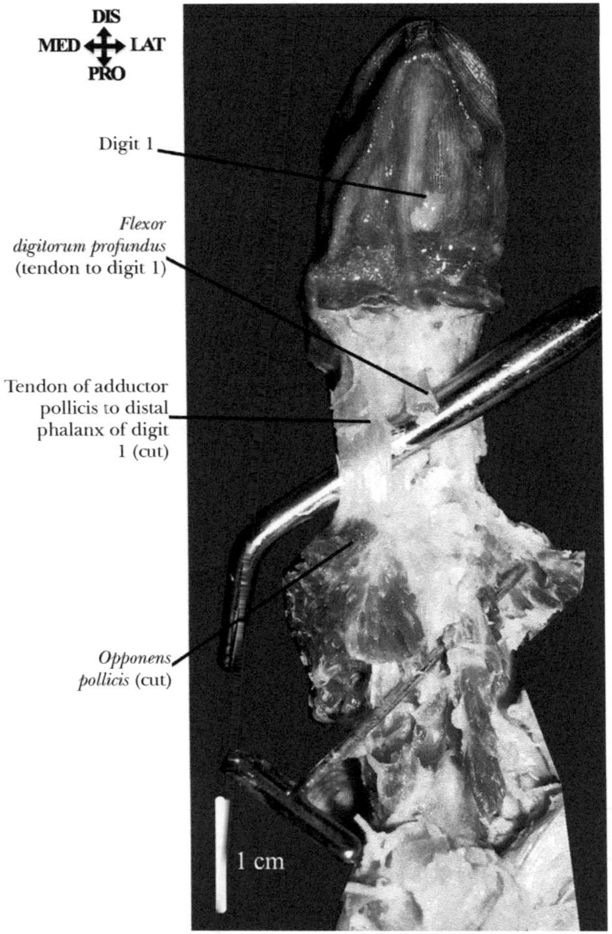

Color Plates 133

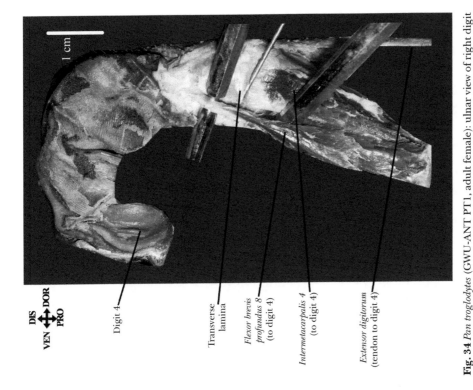

Fig. 34 *Pan troglodytes* (GWU-ANT PT1, adult female): ulnar view of right digit 4 showing the flexor brevis profundus 8 passing superficially to the transverse lamina and the intermetacarpalis 4 passing deep to, or going to, the transverse lamina; contrary to other hominoids, in Pan the flexores breves profundi 3, 5, 6 and 8 and the intermetacarpales 1, 2, 3 and 4 usually do not fuse in order to form the interossei dorsales 1, 2, 3 and 4.

Fig. 33 *Pan troglodytes* (GWU-ANT PT1, adult female): ventral view of the tendons of the flexor digitorum profundus and of the adductor pollicis to the distal phalanx of the right thumb.

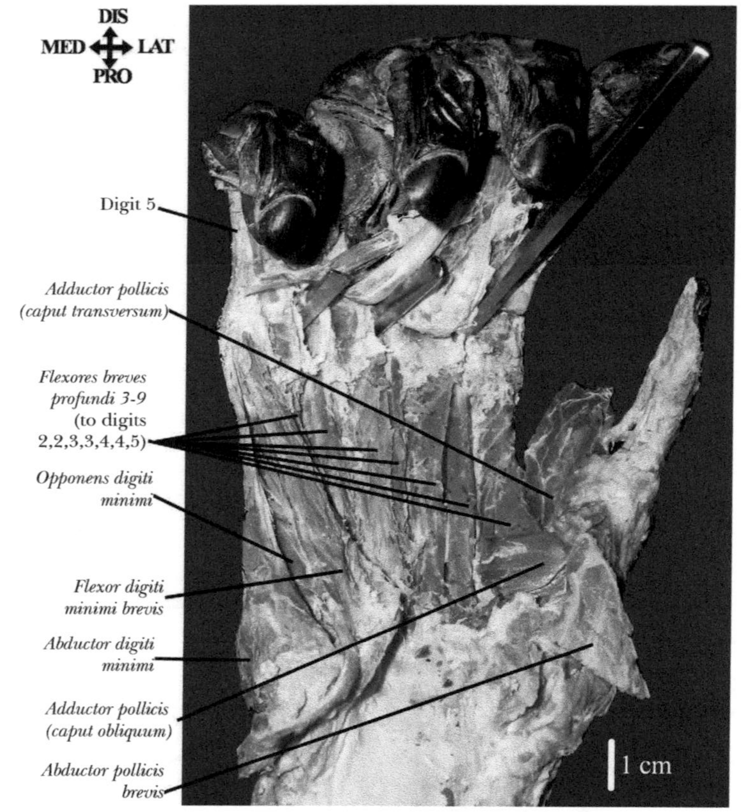

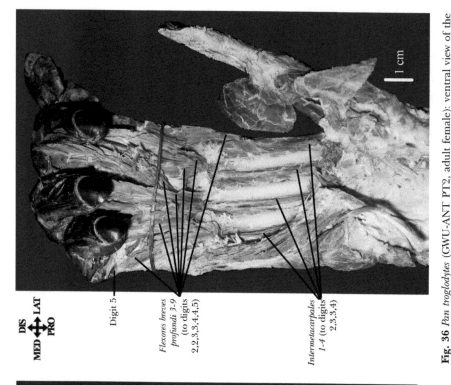

Fig. 35 *Pan troglodytes* (GWU-ANT PT2, adult female): ventral view of the right flexores breves profundi 3-9; contrary to other hominoids, in Pan the flexores breves profundi 3, 5, 6 and 8 and the intermetacarpales 1, 2, 3 and 4 usually do not fuse in order to form the interossei dorsales 1, 2, 3 and 4; the flexores breves profundi 4, 7 and 9 correspond to the interossei palmares of other hominoids.

Fig. 36 *Pan troglodytes* (GWU-ANT PT2, adult female): ventral view of the right flexores breves profundi 3-9 (pulled back) and the intermetacarpales 1-5; contrary to other hominoids, in Pan the flexores breves profundi 3, 5, 6 and 8 and the intermetacarpales 1, 2, 3 and 4 usually do not fuse in order to form the interossei dorsales 1, 2, 3 and 4; the flexores breves profundi 4, 7 and 9 correspond to the interossei palmares of other hominoids.

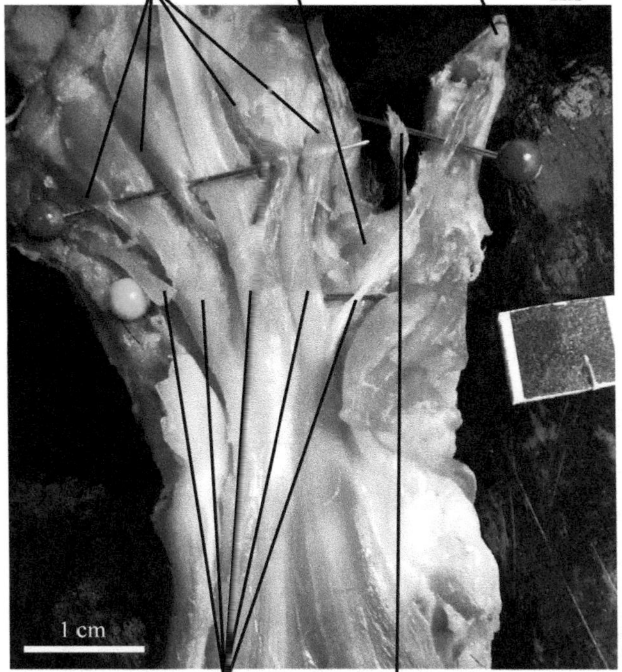

Color Plates 135

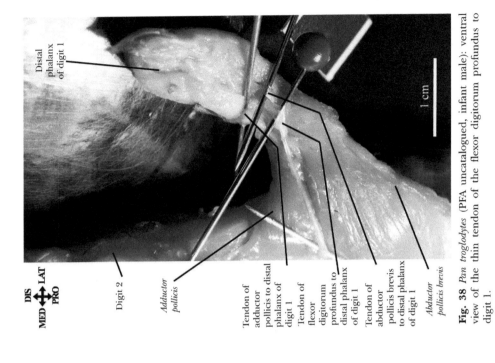

Fig. 38 *Pan troglodytes* (PFA uncatalogued, infant male): ventral view of the thin tendon of the flexor digitorum profundus to digit 1.

Fig. 37 *Pan troglodytes* (PFA 1077, infant female): ventral view of the left flexor digitorum profundus showing the thin tendon of this muscle to digit 1.

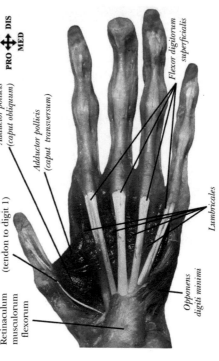
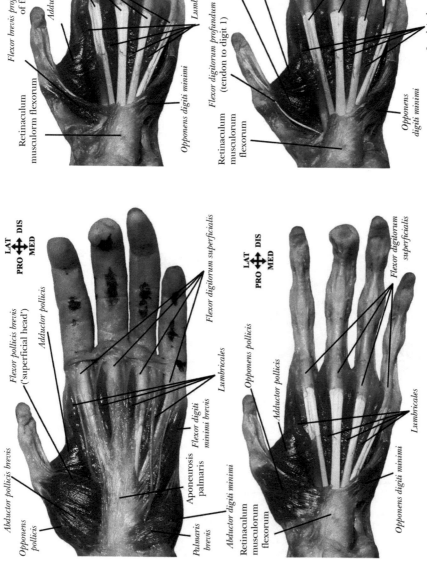

Fig. 39 *Pan troglodytes* (VU PT1, adult male). On the top: palmar view of the superficial muscles of the left hand. On the bottom: same view after removal of aponeurosis palmaris, abductor pollicis brevis, flexor pollicis brevis (superficial head'), abductor digiti minimi and flexor digiti minimi brevis.

Fig. 40 *Pan troglodytes* (VU PT1, adult male). On the top: same view as in Fig. 39 after removal of opponens pollicis. On the bottom: same view after removal of flexor brevis profundus 2.

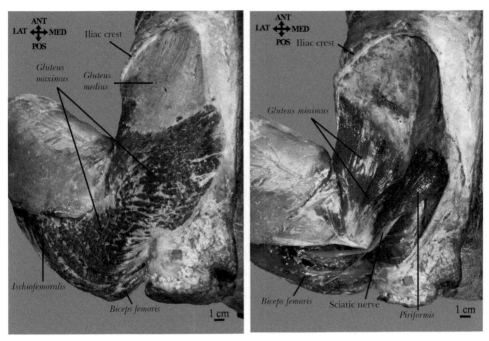

Fig. 41 *Pan trogrodytes* (VU PT1, adult male). On the left: dorsal view of superficial muscles of the left buttock. On the right: same view after removal of superficial muscles of the left buttock.

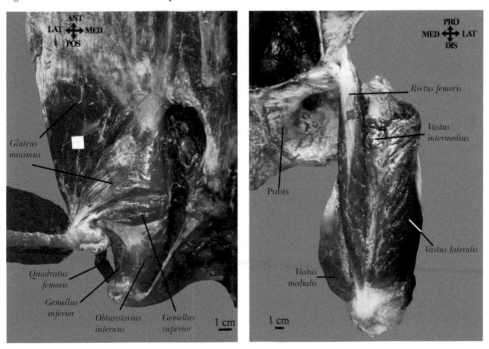

Fig. 42 *Pan trogrodytes* (VU PT1, adult male). On the left: dorsal view of deep muscles of the left buttock. On the right: ventral view of the muscles of the left thigh.

138 Photographic and Descriptive Musculoskeletal Atlas of Chimpanzees

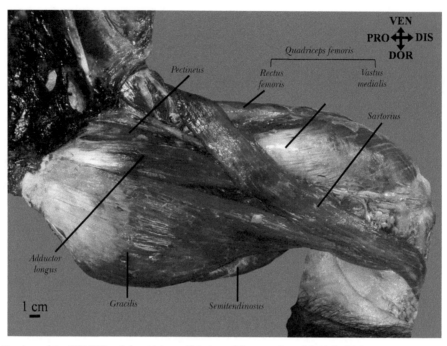

Fig. 43 *Pan trogrodytes* (VU PT1, adult male): medial view of the muscles of the left thigh.

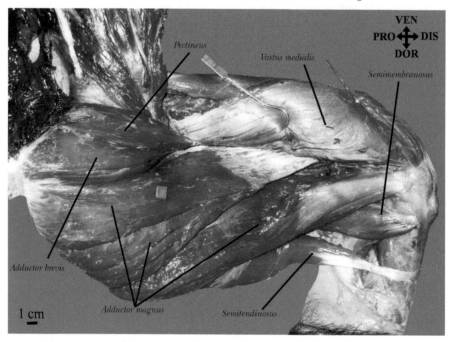

Fig. 44 *Pan trogrodytes* (VU PT1, adult male): medial view of the deep muscles of the left thigh.

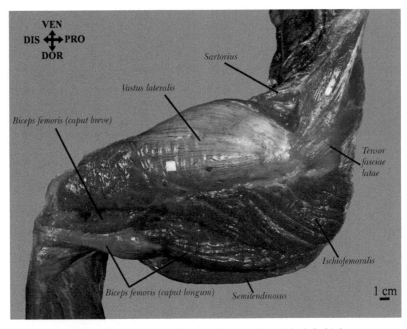

Fig. 45 *Pan trogrodytes* (VU PT1, adult male): lateral view of the muscles of the left thigh.

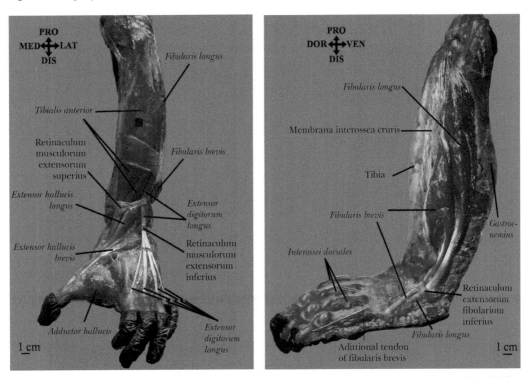

Fig. 46 *Pan trogrodytes* (VU PT1, adult male). On the left: ventral view of the extensor musculature of the left leg and foot. On the right: lateral view of the lateral musculature of the left leg and foot.

140   *Photographic and Descriptive Musculoskeletal Atlas of Chimpanzees*

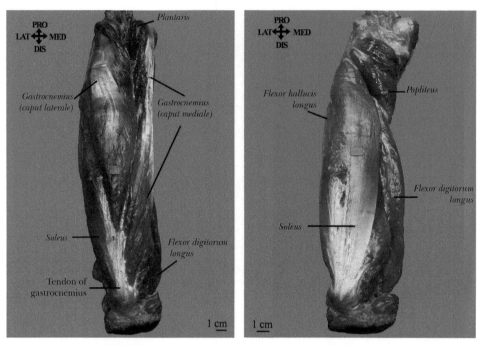

Fig. 47 *Pan trogrodytes* (VU PT1, adult male). On the left: dorsal view of the flexor musculature of the left leg. On the right: same view after removal of gastrocnemius.

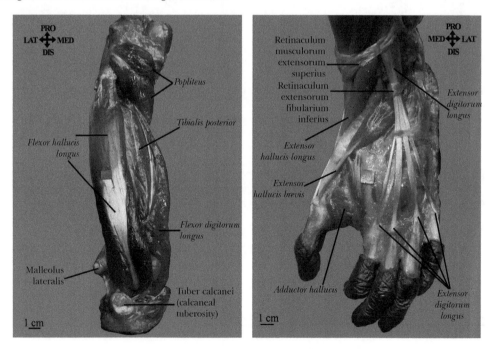

Fig. 48 *Pan troglodytes* (VU PT1, adult male). On the left: dorsal view of the deeper flexor musculature of the left leg. On the right: dorsal view of the superficial flexor musculature of the left foot.

Color Plates   141

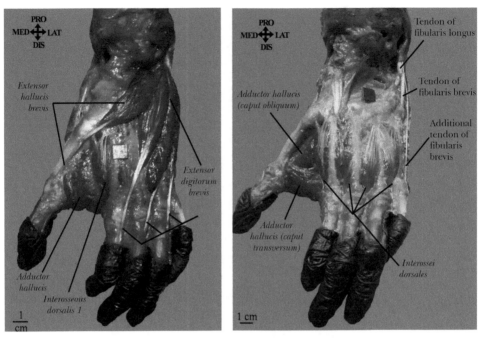

Fig. 49 *Pan trogrodytes* (VU PT1, adult male). On the left: dorsal view of the flexor musculature of the left foot after removal of the most superficial muscles. On the right: dorsal view of the deep flexor musculature of the left foot.

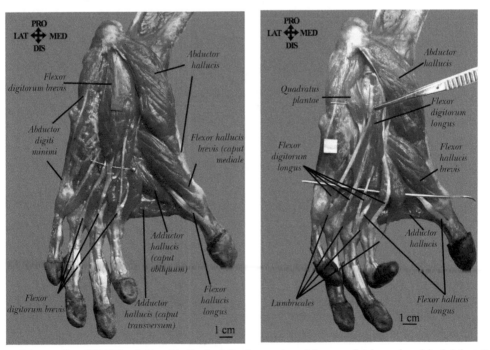

Fig. 50 *Pan trogrodytes* (VU PT1, adult male). On the left: plantar view of the superficial muscles of the left foot. On the right: plantar view of the left foot after removal of the superficial muscles of the foot.

142 Photographic and Descriptive Musculoskeletal Atlas of Chimpanzees

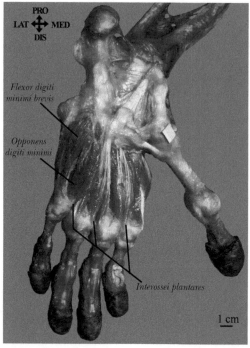

Fig. 51 *Pan trogrodytes* (VU PT1, adult male): plantar view of the deep muscles of the left foot.

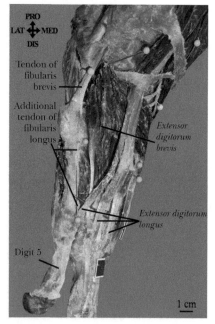

Fig. 52 On the left: *Pan trogrodytes* (VU PT1, adult male): dorsal view of the left thigh, showing an additional belly of the short head of the biceps femoris inserting onto the tendon of insertion of the long head of the biceps femoris. On the right: *Pan trogrodytes* (GWU-ANT PT2, adult female): dorsal view of the right foot muscles, showing an additional tendon of the fibularis longus inserting onto the fifth metatarsal.

Color Plates 143

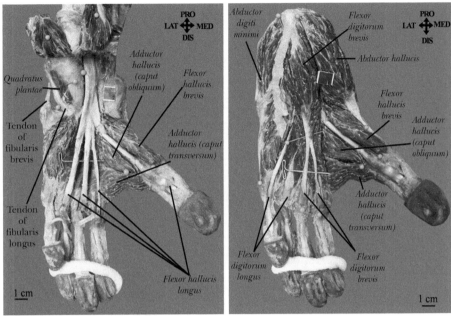

Fig. 53 *Pan trogrodytes* (GWU-ANT PT2, adult female). On the left: plantar view of the muscles of the left foot showing an insertion of the flexor hallucis longus onto the distal phalanges of digits 1, 2, 3 and 4. On the right: plantar view of the muscles of the left foot showing an insertion of the flexor digitorum brevis onto the middle phalanges of digits 2 and 3.

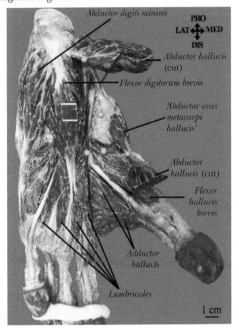
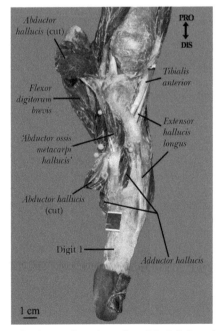

Fig. 54 *Pan trogodytes* (GWU-ANT PT2, adult female). On the left: plantar view of muscles of the left foot, showing a muscle structure that lies deep to the abductor hallucis and that is often designated as 'abductor ossis metacarpi hallucis'. On the right: medial view of the muscles of the left foot, showing the presence of a muscle structure that lies deep to the abductor hallucis and that is often designated as 'abductor ossis metacarpi hallucis'.

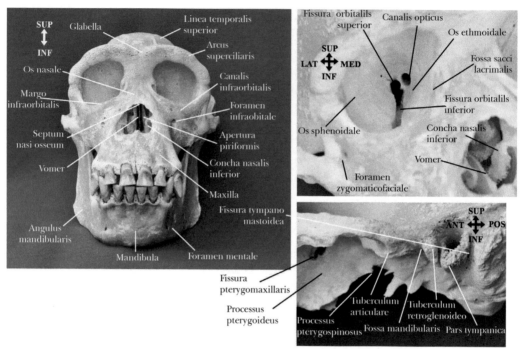

Fig. 55 *Pan trogrodytes* (VU PT3, adult female): frontal view of the cranium (on the left); frontal view of the right orbita (top right); details of the right auditory region and the infratemporal fossa (bottom right).

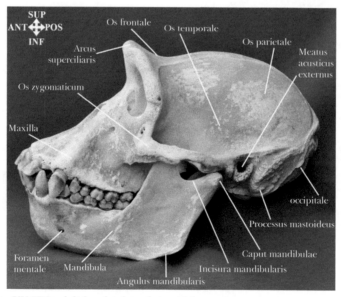

Fig. 56 *Pan troglodytes* (VU PT3, adult female): lateral view of the cranium.

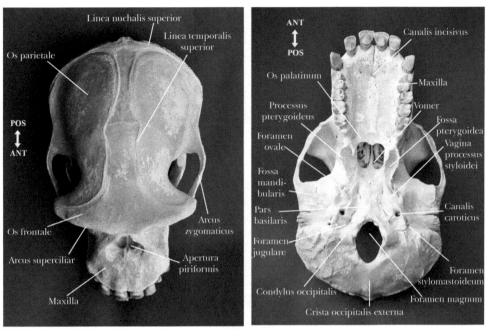

Fig. 57 *Pan troglodytes* (VU PT3, adult female): superior (on the left) and inferior (on the right) views of the cranium.

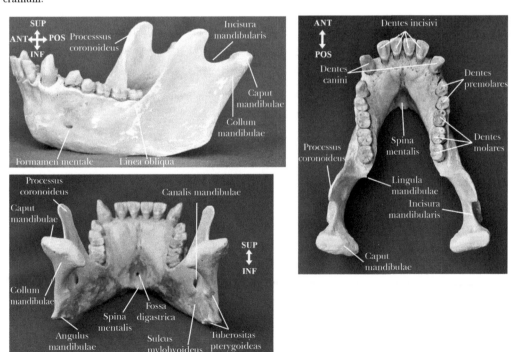

Fig. 58 *Pan troglodytes* (VU PT3, adult female): lateral (top left), superior (right) and posterior (bottom left) views of the mandible.

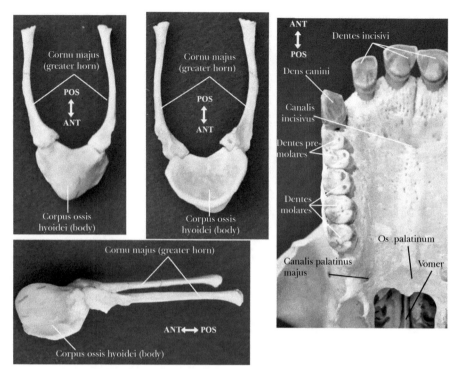

Fig. 59 *Pan troglodytes* (VU PT3, adult female): superior (top left), inferior (top center) and lateral (bottom) views of the hyoid bone; inferior view of the cranium, with details of the maxilla and the palatine bone (top right).

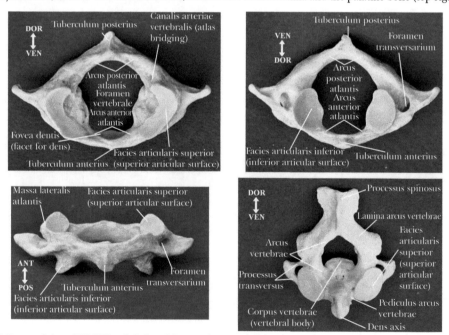

Fig. 60 *Pan troglodytes* (VU PT3, adult female): anterior (top left), posterior (top right) and ventral (bottom left) views of the atlas; anterior view of the axis (bottom right).

Color Plates 147

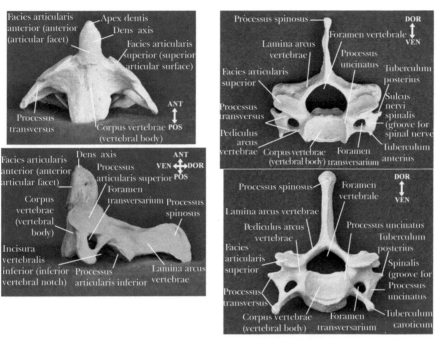

Fig. 61 *Pan troglodytes* (VU PT3, adult female): ventral (top left) and lateral (bottom left) views of the axis; ventral views of a cervical vertebra (top right) and of the sixth cervical vertebra (bottom right).

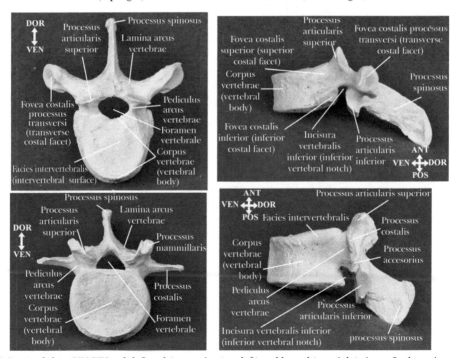

Fig. 62 *Pan troglodytes* (VU PT3, adult female): anterior (top left) and lateral (top right) views of a thoracic vertebra; anterior (bottom left) and lateral (bottom right) views of a lumbar vertebra.

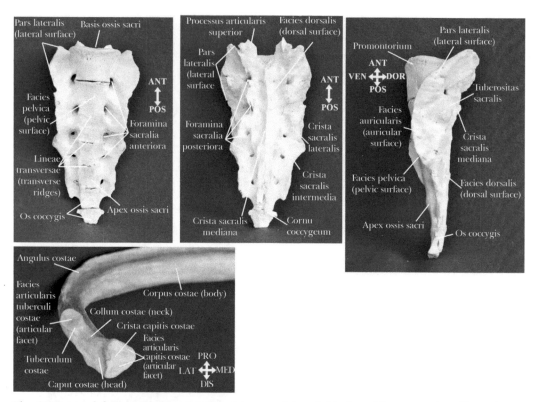

Fig. 63 *Pan troglodytes* (VU PT3, adult female): ventral (top left), dorsal (top center) and lateral (top right) views of the sacrum; dorsal view of the left seventh rib (bottom).

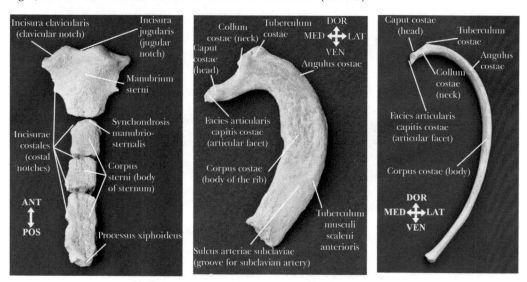

Fig. 64 *Pan troglodytes* (VU PT3, adult female): ventral view of the sternum (on the left); anterior view of the left first rib (on the center); anterior view of the left seventh rib (on the right).

Color Plates 149

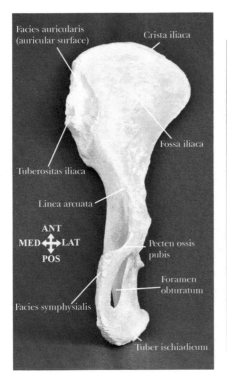
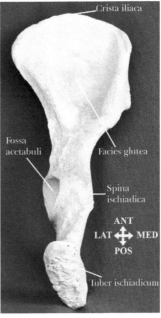

Fig. 65 *Pan troglodytes* (VU PT3, adult female): ventral (on the left) and dorsal (on the right) views of the left coxal bone.

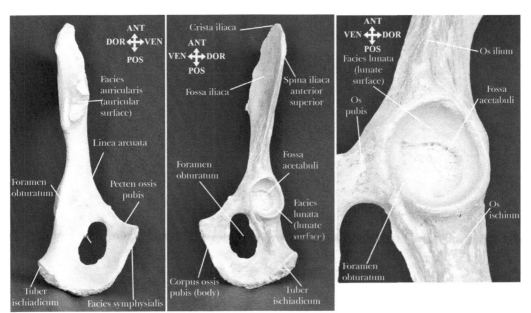

Fig. 66 *Pan troglodytes* (VU PT3, adult female): medial (on the left) and ventrolateral (on the center) views of the left coxal bone and details of the acetabulum (on the right).

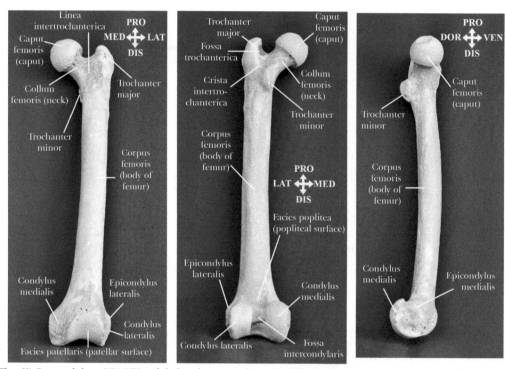

Fig. 67 *Pan troglodytes* (VU PT3, adult female): ventral (on the left), dorsal (on the center) and medial (on the right) views of the left femur.

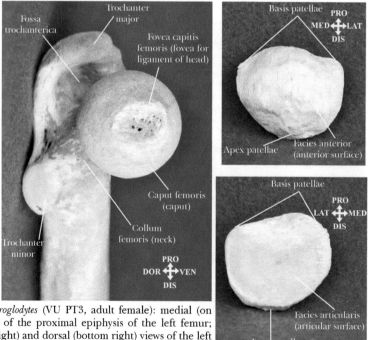

Fig. 68 *Pan troglodytes* (VU PT3, adult female): medial (on the left) view of the proximal epiphysis of the left femur; ventral (top right) and dorsal (bottom right) views of the left patella.

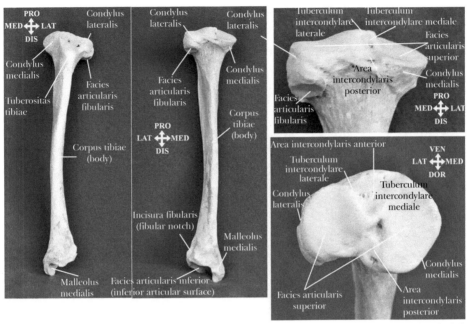

Fig. 69 *Pan troglodytes* (VU PT3, adult female): ventral (on the left) and dorsal (on the center) views of the left tibia; dorsal view of the proximal epiphysis of the left tibia (top right); axial view of the proximal epiphysis of the left tibia (bottom right).

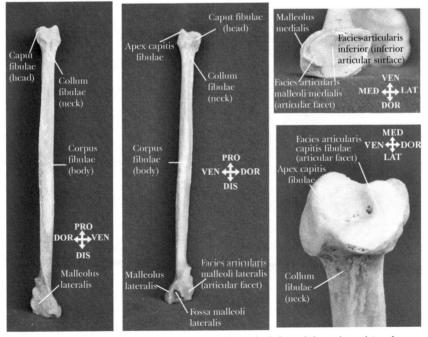

Fig. 70 *Pan troglodytes* (VU PT3, adult female): ventro-medial (on the left) and dorso-lateral (on the center) views of the left fibula; axial view of the proximal epiphysis of the left fibula (bottom right). Axial view of the distal epiphysis of the left tibia (top right).

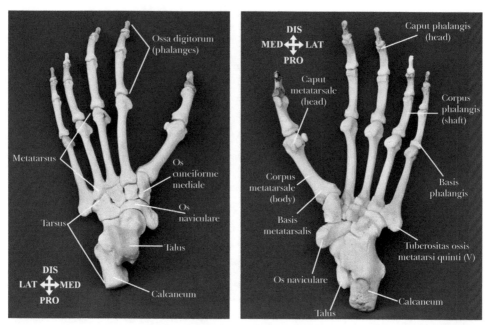

Fig. 71 *Pan troglodytes* (VU PT3, adult female): dorsal (on the left) and plantar (on the right) views of the left foot.

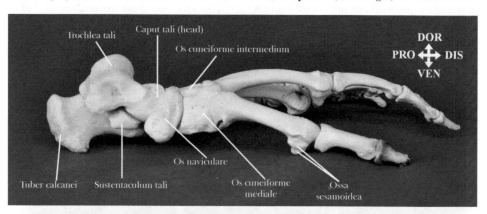

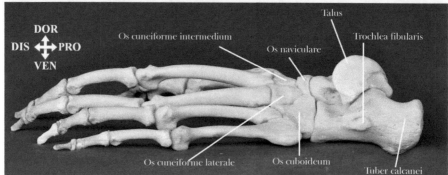

Fig. 72 *Pan troglodytes* (VU PT3, adult female): medial (on the top) and lateral (on the bottom) views of the left foot.

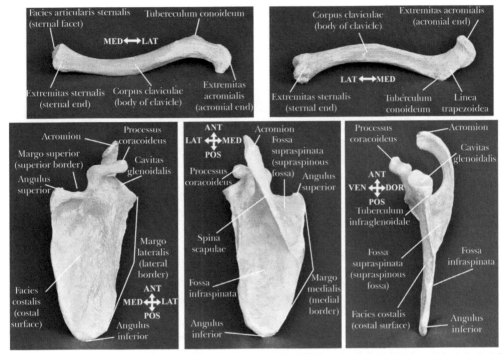

Fig. 73 *Pan troglodytes* (VU PT3, adult female): anterior (top left) and posterior (top right) view of the left clavicle; ventral (bottom left), dorsal (bottom center) and lateral (bottom right) views of the left scapula.

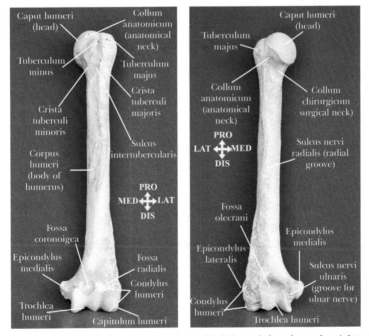

Fig. 74 *Pan troglodytes* (VU PT3, adult female): ventral (on the left) and dorsal (on the right) views of the left humerus.

154 Photographic and Descriptive Musculoskeletal Atlas of Chimpanzees

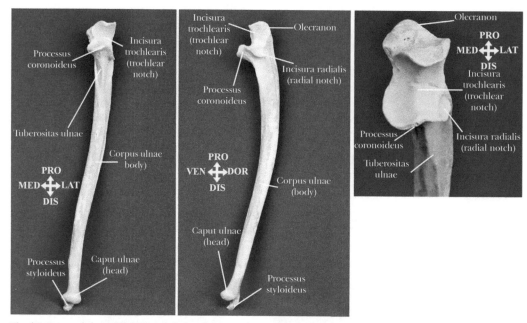

Fig. 75 *Pan troglodytes* (VU PT3, adult female): anterior and lateral views of the left ulna (two figures on the left, respectively); details of the anterior view of the proximal epiphysis of the left ulna (on the right).

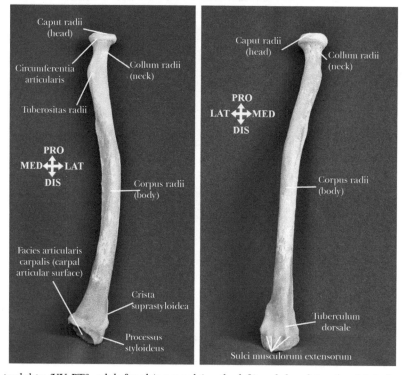

Fig. 76 *Pan troglodytes* (VU PT3, adult female): ventral (on the left) and dorsal (on the right) views of the left radius.

Color Plates 155

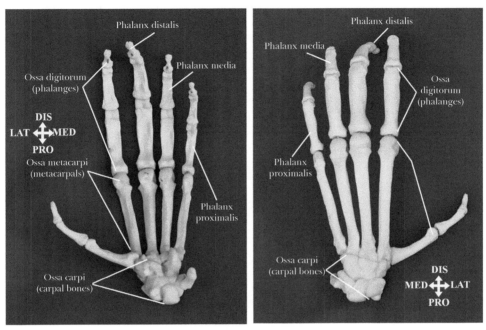

Fig. 77 *Pan troglodytes* (VU PT3, adult female): palmar (on the left) and dorsal (on the right) views of the left hand.

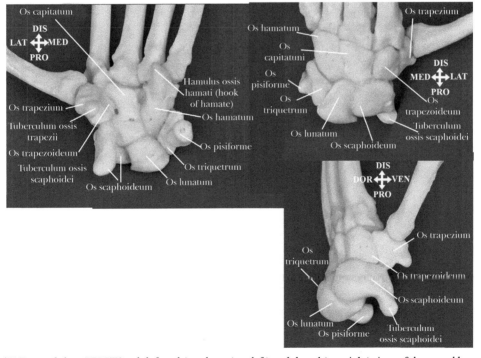

Fig. 78 *Pan troglodytes* (VU PT3, adult female): palmar (top left) and dorsal (top right) views of the carpal bones of the left hand; radial (bottom) view of the carpal bones of the left hand.

For Product Safety Concerns and Information please contact
our EU representative GPSR@taylorandfrancis.com Taylor & Francis
Verlag GmbH, Kaufingerstraße 24, 80331 München, Germany

T - #0178 - 090625 - C168 - 254/178/8 - PB - 9780367380359 - Gloss Lamination